SpringerBriefs in Space Development

Series Editor: Joseph N. Pelton, Jr.

For further volumes:
http://www.springer.com/series/10058

James A. Dator

Social Foundations of Human Space Exploration

Springer

INTERNATIONAL SPACE UNIVERSITY

James A. Dator
Department of Political Science
University of Hawaii at Manoa
Honolulu, HI, USA

ISBN 978-1-4614-3093-3 e-ISBN 978-1-4614-3094-0
DOI 10.1007/978-1-4614-3094-0
Springer New York Dordrecht Heidelberg London

Library of Congress Control Number: 2012930834

Printed on acid-free paper

Springer is part of Springer Science+Business Media (www.springer.com)

ISU (Society) Page

This Springer book is published in collaboration with the International Space University. At its central campus in Strasbourg, France, and at various locations around the world, the ISU provides graduate-level training to the future leaders of the global space community. The university offers a two-month Space Studies Program, a five-week Southern Hemisphere Program, a one-year Executive MBA and a one-year Masters program related to space science, space engineering, systems engineering, space policy and law, business and management, and space and society.

These programs give international graduate students and young space professionals the opportunity to learn while solving complex problems in an intercultural environment. Since its founding in 1987, the International Space University has graduated more than 3,000 students from 100 countries, creating an international network of professionals and leaders. ISU faculty and lecturers from around the world have published hundreds of books and articles on space exploration, applications, science and development.

About the Author

James A. Dator is Professor and Director of the Hawaii Research Center for Futures Studies, Department of Political Science, University of Hawaii at Manoa, Honolulu. He has been Co-Chair of the Space and Society Division of the International Space University, Strasbourg, France, since 1994. He also taught at Rikkyo University (Tokyo, for 6 years), the University of Maryland, Virginia Tech, the University of Toronto, and the InterUniversity Consortium for Postgraduate Studies in Dubrovnik, Yugoslavia.

He is a Danforth Fellow, Woodrow Wilson Fellow, and Fulbright Fellow.

His main areas of interest are political futures studies (especially the forecasting of alternative futures and the design of preferred futures of governance, law, education, and technology); space and society, especially the design of governance systems for space settlements; the political-economic futures of North America, the Pacific Island region, and east Asia, especially Japan and Korea; and media production and the politics of media, and the effects of media on political and other human relations and consciousness.

He has consulted widely all over the world in these areas and is in demand as a public speaker.

Email: dator@hawaii.edu
Website: www.futures.hawaii.edu

Contents

Chapter 1
What Is This Book About?

This book presents a uniquely human perspective on the quest to explore space and to understand the mysteries of the universe as seen through the lens of the arts and humanities, the social sciences, history, and futures studies. The contribution that the social sciences and the humanities have been allowed to make to space studies and actual space endeavors to date is very slight. In fact the contribution of these various disciplines is modest when compared to that of science, engineering, medicine, law, policy, and business. Yet, the systematic, early, and omnipresent inclusion of social science research and humanistic concerns into all future space activities is vital for the success of any space mission. Indeed, one of the best justifications for the international space station (ISS) – for example – should have been to perform good social science and humanistic research in preparation for space flights and settlements of longer duration elsewhere in the Solar System. And yet no such research has been conducted on the ISS. In contrast scores of biological, mechanical, and astronautical experiments have been carried out, while social science and humanistic research has largely been considered beyond the purview of most space agencies. It should be noted, however, that JAXA, the Japanese Space Agency, does have a series of poems called "renshi" written by ordinary people around the world on the *Kibo* module of ISS.

The Focus of This Book

This book argues that a diversity of social sciences, arts, literature, and humanities are indeed quite relevant to space. In the chapters that follow we explore a number of the "hows" and "whys" of these relationships. It is simply not possible to cover all of the aspects of space and society in this quite brief book. This is a primer – a basic introduction to the academic study of space and society.

We will start by addressing art and literature over time and in many cultures in all of its many forms as it relates to space. Then we will discuss the historical origins

J.A. Dator, *Social Foundations of Human Space Exploration*, SpringerBriefs in Space Development, DOI 10.1007/978-1-4614-3094-0_1, © James A. Dator 2012

of what might be called "the space age." Next we consider the cultural rationale for space (in contrast to political, economic, scientific, or military rationales). We will provide some brief but key observations describing the experiences of the few humans who have been into space – that is, to the Moon and in Earth-orbit. Another chapter will probe the coming experiences in space involving advancing technology. This chapter will address possible futures of humans in space with robots, artilects (i.e. artificial intellects) and transhumans. The next chapter will explore ethical considerations of space activities, including whether "rocks have rights." In the final three chapters we will examine the search for extraterrestrial intelligence, designing governance systems for space, and alternative futures for space exploration and settlement. In short we will range far and wide in the pages that follow and we hope you enjoy the ride.

Consideration of the humanities and social sciences in space has not been entirely neglected in the academic world. One part of the space community that has embraced the humanities fully into its space-related activities is the International Space University (ISU) headquartered in Strasbourg, France. ISU was conceived and founded in the late 1980s by three young men – Todd Hawley, Bob Richards, and Peter Diamandis – who were then graduate students. Their "Credo" for ISU, proclaimed in April 1995, inspiringly states:

WE, THE FOUNDERS of the International Space University, do hereby set forth this Credo as the basis for fulfilling ISU's goals and full potential.

INTERNATIONAL SPACE UNIVERSITY is an institution founded on the vision of a peaceful, prosperous and boundless future through the study, exploration and development of Space for the benefit of all humanity.

ISU is an institution dedicated to international cooperation, collaboration and open, scholarly pursuits related to outer space exploration and development. It is a place where students and faculty from all backgrounds are welcomed; where diversity of culture, philosophy, lifestyle, training and opinion are honored and nurtured.

ISU is an institution which recognizes the importance of interdisciplinary studies for the successful exploration and development of space. ISU strives to promote an understanding and appreciation of the Cosmos through the constant evolution of new programs and curricula in relevant areas of study. To this end, ISU will be augmented by an expanding base of campus facilities, networks and affiliations both on and off the Earth.

ISU is an institution dedicated to the development of the human species, the preservation of its home planet, the increase of knowledge, the rational utilization of the vast resources of the Cosmos, and the sanctity of Life in all terrestrial and extraterrestrial manifestations. ISU is a place where students and scholars seek to understand the mysteries of the Cosmos and apply their knowledge to the betterment of the human condition. It is the objective of ISU to be an integral part of Humanity's movement into the Cosmos, and to carry forth all the principles and philosophies embodied in this Credo.

THIS, THEN, IS THE CREDO OF ISU. For all who join ISU, we welcome you to a new and growing family. It is hoped that each of you, as leaders of industry, academia and government will work together to fulfill the goals set forth herein. Together, we shall aspire to the Stars with wisdom, vision and effort.

On the basis of this commitment, ISU activities in the humanities and social sciences seek to apply the traditional and evolving perspectives of all of the various academic disciplines of the arts, humanities, and social sciences to space-related activities both on Earth and in space, and to space-Earth interactions.

Probing Beyond

The Space and Society (sometimes called Space Humanities) is especially responsible for actively developing, freely examining, and rigorously critiquing the various historical, futures-oriented, philosophical, ethical, esthetic, spiritual, and social issues and perspectives concerning space exploration, and humanity's place on Earth and in the cosmos It strives to cover all of the fine arts, humanities, and social sciences as they relate to space. This is a gigantic task. Clearly we cannot cover all of those topics in this book, but our aim is to try to embrace them all – if but briefly – in some way.

Interdisciplinary Issues

One major problem is that academia is currently divided into departments that make it difficult to consider interdisciplinary issues. The real world – certainly the world of space – is increasingly international, intercultural, and interdisciplinary, and not divided into the neat boxes of academia. Each of the academic disciplines came into existence at a certain time for certain reasons. Some of those reasons are still completely valid, but others may no longer be valid at all.

In sum, this book uses the perspectives of the social sciences and humanities to understand and/or guide:

1. *Space-related activities on Earth*
   This exploration includes the history of human ideas about space, and the expression of those ideas including space art; history of human space exploration; opposition of some to space exploration; space education and outreach.
2. *Space-related activities in space*
   Such investigations include the governance systems of space settlements; psychology and sociology of human space exploration; art in space; human interaction in space environments; human-artificial intelligence interactions in space.

3. *Space-Earth interactions*
   These interactions include conflicts between mission ground control and space crews; search for extraterrestrial life; encounters with unidentified flying objects; evolution of spacekind from Earthkind; and what has been called cosmicization.

Contributors to This Book

Every human who has ever thought about, worked for, or lived in space has also engaged in some aspect of this topic simply by virtue of being human and bringing his or her passions, prejudices, fears and hopes to the enterprise. Moreover many people educated as scientists, engineers, mathematicians, medical doctors, business managers, communications experts, satellite operators, and so on have made major contributions to the literature and field of space exploration. They are simply too numerous to list, and any listing would almost certainly unintentionally omit many who deserve recognition. But the number of people who are formally trained in the social sciences or humanities and have applied their expertise in some major way to developing the field of space development is very small indeed. In that regard, I pay my special respects to Jacques Arnould, Linda Billings, Sheryl Bishop, B. J. Bluth, William Burrows, T. Stephen Cheston, Steven Dick, Kerri Doughtery Jim Funaro, Philip Harris, Albert Harrison, Todd Hawley, Nick Kanas, Roger D. Launius, John Logsdon, William MacDaniel, Jim Pass, Joe Pelton, George Robinson, Patricia Santy, Ellwyn Stoddard, David Swift, Allen Tough, Douglas Vakoch, Stewart Whitney, and Tsutomu Yamanaka. Without their guidance, the special mentoring of Ben Finney, and the continuing encouragement of Jim and Lin Burke, I would not have been able to offer this slight contribution to the field.

References

T.S. Cheston et al. (eds.), *Social Sciences and Space Exploration: New Directions for University Instruction* (US Government Printing Office, Washington, DC, 1984), NASA EP192

A. Houston, M. Rycroft (eds.), *Keys to Space: An Interdisciplinary Approach to Space Studies* (McGraw-Hill, New York, 1999)

J.N. Pelton, A. Bukley (eds.), *The Farthest Shore: A 21st Century Guide to Space* (Apogee Books, Burlington, 2010)

Chapter 2
Visions of Not-Earth: Space Arts

Dreams of space – of not-Earth – have inspired humanity over the ages. Dream-inspired humans made space travel a reality. Without our dreams, there would be no space programs anywhere. Artistic, religious, philosophical, and ethical perspectives are not frills or mere add-ons to space activities. They are absolutely essential parts of all aspects of all space endeavors. At the same time, without the science and technology that enables humans to loose the bonds of Earth, humans would still only be dreaming of space while never going to the Moon and beyond. But science and technology are themselves the products of human dreams and desires. Dreams, beliefs, science, and technology – along with natural and human resources harnessed by human will and labor – are all required to attain and maintain space activities.

Stories of Not-Earth

Every culture on Earth has some kind of foundational explanation for the creation of Earth and of humans. Many of these stories say that humans, other life, and Earth were created either by nonhumans or by super humans who came down from the sky. There are virtually no creation stories that clearly follow the perspective of Darwinian evolution although sometimes people try to show that they do. Some, such as the Abrahamic tradition, may have God creating heaven and Earth in 7 days, which some try to interpret now in evolutionary terms. However, the message is not one of natural selection with unguided evolutionary development over eons of time, but of creation by an external sentient being.

Many cultures also tell of voyages between Earth and not-Earth. The idea that there are worlds and beings not of this Earth that interact with humans and Earth is very widespread across the globe and has persisted for a very long time. Modern space fiction is just one current way by which very old stories are being told and retold dressed up with contemporary ideas and technologies. These old stories influence the way we think about current space exploration.

J.A. Dator, *Social Foundations of Human Space Exploration*, SpringerBriefs in Space Development, DOI 10.1007/978-1-4614-3094-0_2, © James A. Dator 2012

For example, one of the best-known Japanese folktales is about Kaguya Hime. A bamboo cutter discovered a baby girl inside a bamboo shoot. He took the baby home, where he and his wife reared the baby as their own. When she grew up, she revealed that she was not from Earth and was transported back to the Moon from which she came. Her story, and her return to the Moon, is repeatedly told in Japan. In the summer of 2007, JAXA (the Japanese space agency) launched a lunar orbiter. JAXA asked the Japanese public what the name of the orbiter should be, and Kaguya Hime was the overwhelming choice.

Another popular folktale from China, Korea, and Japan is told every summer on the seventh night of the seventh month. It celebrates what appears to be the annual meeting of the stars Vega and Altair that are otherwise separated by the Milky Way. According to the folk story, Vega is a seamstress and Altair is an ox-herder. They love each other but are separated from one another and only allowed to meet briefly once a year, and then must part once again.

The Japanese space agency named a pair of satellites *Orihime* (the seamstress) and *Hikoboshi* (the ox-herder). The two were launched together as ETS7 and then separated before coming together again in 1999. It was the first time automated docking had ever been carried out in space. Most of the vehicles that Japan has sent into space have names from folktales, or are named such things as *Kibo* (Hope), *Kizuna* (Wind), *Kiku* (Chrysanthemum, the national flower), *Hinode* (Sunrise), and *Kodama* (Spirit).

Compare them to the names NASA has given to some of its space vehicles: Mercury, Apollo, Gemini, Saturn, Jupiter, Orion, Titan. They are all named after mighty Greek or Roman gods with fearful powers, conveying a very different impression in comparison with the Japanese terms from nature or folktales.

However, probably the iconic figure for flight in Western cultures is Icarus, the Greek who, with his father, Daedalus, fashioned wings so that he and his father could fly like the birds, something they were totally unable to do with their wingless, flightless natural bodies. So, as is always the case, humans first imagine doing something impossible – flying – and then develop the technologies that enable their dreams to come true. However, Icarus also received a warning from his father before he took off. According to Ovid, in one English translation, Daedalus said:

> My Icarus, I vow thee fly
> Always the middle track; nor low, nor high;
> If low, thy plumes may flag with ocean's spray
> If high, the sun may dart his fiery ray.

Of course Icarus, being a typical son – and a typical human – disregarded his father's warning. He did fly. But then he sailed too close to the Sun, which melted the wax on his wings and cast Icarus to his death in the ocean below. The daring and hubris of Icarus has been an extremely popular theme in Western art and literature, warning us of the eternal tension between what we *want* to do and then *can* do because we develop technological capabilities, on the one hand, in contrast with what we *ought* to do, given our ethical limitations and frailties, on the other. The Icarus story has had enormous impact on Western consciousness and is depicted in many works of art and poetry as an example of humanity's daring spirit, as well as our hubris and limitations.

Voyages of Discovery

Space fiction, almost by definition, involves voyages of discovery. However, well before the modern era, certain cultures had stories about voyages of discovery, while other cultures had no such stories at all. In the former, heroes leave home, travel through strange times and places, overcome many adversities and have many exceptional experiences before returning home again, enlightened by the process. The basic archetypical stories for Western cultures are the *Iliad* and the *Odyssey,* first composed between 800 and 600 BCE. The *Odyssey,* recounting 20 years of travel by Odysseus (Ulysses, in Latin), is a prime example.

In the Abrahamic Bible, the first book, Genesis, is immediately followed by Exodus: departure happens soon after Creation. The story of Moses leading the Jewish people to the Promised Land – and the belief in the existence of a Promised Land that is rightfully theirs – is an unfinished narrative of travel and travail in Western cultures. In Christian belief, the Wise Men traveled far to find the Messiah. The Muslim faithful must travel to Mecca. There appears to be an almost irresistible urge in certain cultures for humans Boldly To Go – or at least for some people, usually men, to go. In each of the stories above there are those who warn against the journey, and/or who patiently stay at home waiting for the hero's eventual return, such as Ulysses' Penelope.

The stories and actual experiences of the people who left what is now south China perhaps 4,000 years ago and spread across the vast Pacific Ocean from island to island, creating the very diverse cultural groups known as Melanesia, Micronesia and Polynesia, and stopping only when they finally landed in Hawaii 3,000 years later is an extraordinary episode of actual human voyaging. They went on their dangerous voyages in part perhaps because they were forced to move by population pressure and resource scarcity, but perhaps mainly because of their cultural stories that compelled them to travel, and the unique technologies they developed that enabled them to navigate by the stars, winds, clouds, tides, and fish movements in order to go where no one had ever gone before.

Stories and examples of heroic travel to unknown places do not exist in some cultures. They instead are told in effect to stay home, to get along with their neighbors, and to mind their own business. The urge boldly to go is found in some humans and cultures, but by no means in all. In those cultures where exploring is deeply rooted in myths, "going boldly" is almost irresistible, while in others with no such stories, it seems almost impossible to ignite.

Emerging Science Fiction and a Sense of the Future

Around A.D. 150, the Greek philosopher Lucian of Samosata wrote what might be the first two Western works of space fiction, *Icaromenippus* and *The True History.* In the first, Menippus (specifically wanting to avoid the failure of Icarus) took one wing from an eagle and another from a vulture and fashioned them so he could fly from Mount Olympus to the Moon. In Lucian's second story, a ship exploring the Atlantic was carried by a waterspout to the Moon.

Lucian's stories are more than mere fanciful tales. In them, Lucian ridiculed humanity, its tired old philosophies and pagan beliefs, and the emptiness of the intellectual life of the time – all themes that we find repeatedly in space fiction.

Though written as theological works, St. Augustine's *De Civitate Dei* in the fifth century, and Joachim of Fiore's *Liber Concordiae Novi ac Veteris Testamenti* and *Expositio in Apocalipsim* in the early thirteenth century, each exhibited futures-oriented utopian thinking. The great Muslim scholar, Ibn Kahldun, writing in the fourteenth century, is considered one of the fathers of sociology and of futures studies, presenting a sophisticated philosophy of history and society in his *Muqaddimah* – not a work of fantasy or fiction, but rather exhibiting, perhaps for the first time, a way of thinking about humans and their past and future that greatly influenced the emergence of modern attitudes towards social change, and hence, modern space fiction.

Early Science Fiction

Science fiction is seldom fiction about science. Rather, it is stories that arise when some people become aware of the fact and possibility of continuous social change. Thus, science fiction *per se* (and hence space fiction) is a product of the scientific, technological, and industrial revolution that was made possible in Europe by the Black Plague, the Reformation, and the Renaissance between the fourteenth and seventeenth centuries, and then bloomed during the late eighteenth, nineteenth, and twentieth centuries. In the mid nineteenth century, space fiction proper emerged first in mainland Europe, then in the UK, then simultaneously in the United States, Japan, China, India (and perhaps elsewhere) as the social and environmental consequences of the scientific, technological and industrial revolution spread across the globe.

Most science fiction is more about technology than it is about science. It is about how humans might behave and how society might change if/as new technologies come along. Often the "science" in science fiction is quite unscientific, though the behavior said to result from new technologies is sometimes more plausible. But much science fiction is bad social science as well as bad natural science – and not very good fiction either. Science fiction and space fiction thus exhibit a tension between two modes of knowing – one scientific, the other fictional. Some science fiction is closer to science than to fiction, and thus often boring though factual. Most science fiction is closer to fiction than to science, thus exciting but misleading. The best science fiction finds a balance between both.

Modern Science Fiction and Actual Spaceflight

The literature of science fiction and space fiction dealing with new technologies and technological change has generally been of one of two kinds. Jules Verne and many others were basically optimists, believing in inevitable progress through technological change.

This optimistic view of the future permeated much early science fiction and space fiction. But from the beginning, other science fiction writers had more of a love-hate relation with technology and often wrote of that relationship critically and sardonically.

Without a doubt, the most important single figure in the origins of science fiction and space fiction is the French author, Jules Verne. Verne's book, *De la Terre à la Lune* (From the Earth to Moon, 1865), and many others of his books were translated into every major language of the world. During his lifetime Verne was perhaps the most widely read author in the world, and his books are still popular. Almost all early pioneers in space reality and space fiction said that they were inspired by Verne.

The importance of space fiction in creating space reality – and vice versa – cannot be overstated.

The extraordinary pioneer of Russian spaceflight and space fiction, Konstantin Tsiolkovsky, said his enthusiasm for space came from reading Jules Verne. In addition to his vital role in envisioning and enabling actual spaceflight, Tsiolkovsky himself wrote classics of Russian science fiction including one that the world's first cosmonaut, Yuri Gagarin, said was his favorite: *Vne Zemli* (*Beyond the Earth*, 1896). An enormous amount of space fiction produced in the Soviet Union was intended to inspire that nation towards new futures.

Although the earliest science and space fiction originated in Europe, there were clear precursors in other countries. Science and space fiction *per se* emerged rapidly wherever industrialization and technologically induced social change spread in the world.

The roots of science fiction in India may stem from 1500 BCE in the ancient Vedic literature. In these texts there are descriptions of what some say are unidentified flying objects, referred to as *vimanas*. However, science fiction proper emerged in India when, as one author says, "the effects of the industrial revolution were being felt in urban India in the nineteenth century just as keenly as they were in Europe and the U.S." The earliest notable Bengali space fiction was Jagadananda Roy's *Shukra Bhraman* ("Travels to Jupiter"), written in 1857. This story is of particular interest as it described a journey to another planet, while the existence of the creatures seen there was explained using evolutionary concepts. It should be noted that this story was published well before H. G. Well's *The War of the Worlds* (1898) in which Wells described an invasion from Mars.

The Indian space agencies first space satellite, in 1975, was named *Aryabhatta,* after the revered fourth-century mathematician. India's first satellite for Earth observation, launched in 1979, was named *Bhaskara,* after the famous twelfth-century mathematician and astronomer.

Chinese creation stories typically have themes involving space, the cosmos, and chaos, such as Pan Gu, who separated the heavens from Earth, and Nu Wa, who patched up the falling heavens. Similar themes are found in the earliest Chinese literature, such as *Chang E benyue* (Chang E Goes to the Moon) by Lu An (197-122BCE), which is about Chang E, who was able to fly to and live on the Moon. The first Chinese lunar orbiter, launched in October 2007, was named *Chang'e-1.*

Henry Zhao states that "prior to the concept of modernity being imported into China there had been no fiction about the future. In traditional China, history did not have directionality." The introduction first of ideas about "progress" and "development" and then of Marxism changed that. The young intellectuals of the late nineteenth century in China sought to bring their "backward" country into a modern nation-state. To do that – they learned from Japan and the West – science and technology was necessary. So to stimulate people's interest in science and technology, Lu Xun introduced science fiction to China with his 1903 translation of Jules Verne's novel *From the Earth to the Moon.* His translation is posted on the official website devoted to the *Chang'e-1* lunar orbiter.

The mission of Chinese science fiction has always been to encourage people to appreciate the power of science and technology in enabling their country to develop into a modern nation. Chinese science fiction is optimistic and utilitarian rather than simply aesthetic.

We have seen that some Japanese folktales were explicitly about creatures who came to the Earth from not-Earth and eventually returned. Translations of Verne and other writers were extremely popular in Japan from the 1880s onward, and there was an explosion of science and space fiction after the Second World War. Nowhere has space fiction been more popular than in Japan.

H. G. Wells is to English science and space fiction what Jules Verne was to European – and the world's – science fiction. Among Wells' best-known stories are *The Time Machine* (1895), *The War of the Worlds* (1898), and *The First Men in the Moon* (1901).

In the United States, science's and space fiction's heyday is found in the "pulp" magazine, *Amazing Stories*, that began publication in 1926, and *Astounding Stories* that went through many name changes, ending up as *Analog* today. Many other pulp magazines devoted to science and space fiction existed from the 1930s to 1950s, and almost all the great and not so great names of science fiction history published in them, defining the genre from that point onward.

Not-Earth in Movies, Television, and Games

Movies, and later, television, have almost always featured some kind of "space opera," from *Le Voyage dan la Lune*, ("A Trip to the Moon," 1902) through *Buck Rogers* and *Flash Gordon* in the 1930s and 1940s, to *Tetsuwan Atomu* after the Second World War, and many others onward. Even though most of these films were "B" grade movies at best, they and the pulp science fiction books of the era created themes that defined and have persisted in almost all space fiction everywhere in the world.

Stanley Kubrick's co-production with Arthur C. Clarke of *2001: A Space Odyssey* (1968) brought space fiction in films to a high level, while Roger Vadim's adaptation of the French comic strip character *Barbarella* (1968) was a magnificent high camp film in the ironic mode. There is no more striking contrast in space films than between *2001*s cold, barren, functional spaceship (with the onboard intelligent computer, HAL, and the human Dave locked in a gripping battle of wits to the death) and the naïve, nubile, and naked Jane Fonda slithering across her fur-lined spaceship.

On television, the quirky British space fiction show *Dr. Who* (1963 onward even to the present!) influenced many viewers' ideas about space. But for Americans, good space fiction on television began with the extremely popular TV series *Star Trek* in 1966, which dealt with current social, political, and ethical issues in the guise of exploring the universe. After the blockbuster movie *Star Wars* in 1977 (and its successors), space fiction on television for some time was dominated by triumphal voyages of international, intercultural, and interdisciplinary members of a united Earth federation boldly exploring the cosmos, doing good (or at least not doing evil) while trying to obey "the prime directive" of not interfering in the lives of other cultures.

One of the most important developments in science fiction was the emergence of cyberpunk literature spearheaded by William Gibson's *Neuromance*r (1984). It was inspired by contemporary and emerging advances in electronic communication technologies, biotechnology, and nanotechnology, combined with deep anxiety about the environmental and social consequences of these and other developments. Cyberpunk treats science and technology critically and ironically. A rather Gothic form of cyberpunk pervades most interactive electronic games, a form of science and space fiction that may replace not only written literature but also cinema and television. Contemporary Japanese science fiction, especially in its *anime* and *manga* forms, is heavily cyberpunkish.

So far, we have concentrated entirely on space stories in print and modern media. But space has inspired many works of poetry, drawing, painting, sculpture, pottery, weaving, music, and dance.

Poetry About Not-Earth

There are countless poems and stories that humans have told upon looking up at the dark sky at night. But as scientific ways of thinking were beginning to challenge earlier cultural modes, art in all its forms was also influenced by the new ideas, technologies, and discoveries. For example, William Drummond wrote *The Shadow of the Judgment* at a time (1630) when the discovery of new stars was viewed as an omen of the end of the world, since (according to the philosophical and religious thought of the time) the heavens should be fixed and unchanging. The telescope also led to much speculation about life elsewhere in the universe – and the insignificance of the petty squabbles of humans on Earth in comparison.

By the twentieth century, Einstein, Heisenberg, Schrodinger, Plank and others in physics, and Eddington, Wheeler, Bell, Penrose and others in astronomy were putting the earlier physics of Newton and the astronomy of Brahe, Galileo, et al into a different light. Alfred Noyes expresses this very well:

> "In the time to come,"/Said Tycho Brahe, "perhaps a hundred years, / Perhaps a thousand, when our own poor names / Are quite forgotten, and our kingdoms dust, / On one sure certain day, the torch-bearers / Will, at some point of contact, see a light / Moving upon this chaos. Though our eyes / Be shut forever in an iron sleep, / Their eyes shall see the kingdom of the law, / Our undiscovered cosmos. They shall see it, – / A new creation rising from the deep, / Beautiful, whole.

> We are like men that hear / Disjointed notes of some supernal choir. / Year after year, we patiently record / All we can gather. In that far-off time, / A people that we have not known shall hear them, / Moving like music to a single end." [From *Watchers of the Sky*, II. *Tycho Brahe*, by Alfred Noyes (1922)]

Until 1957, no one had actually been in space (except on spaceship Earth!) and so all poets were writing from imagination, not from their own direct experience. But there have been poems written by astronauts and cosmonauts. During the nearly 67 h his fellow astronauts, Scott and Irwin, were on the moon during the *Apollo 15* mission, Al Worden was in complete solitude, floating in space. He said the overwhelming experience of being alone in the universe gave him a profound feeling of rejuvenation. In 1974 Worden wrote a book of poems entitled, *Hello Earth: Greetings from Endeavor*. As for so many others, Worden's experience in space changed his view of reality on Earth:

> Quietly, like a night bird, floating, soaring, wingless / We glide from shore to shore, curving and falling / but not quite touching; / Earth: a distant memory seen in an instant of repose, / crescent shaped, ethereal, beautiful, / I wonder which part is home, but I know it doesn't matter … / the bond is there in my mind and memory; / Earth: a small, bubbly balloon hanging delicately / in the nothingness of space.

Another American astronaut, Story Musgrave, has written a great deal of poetry based on his own experiences.

Floating in a spaceship,
Falling through my heaven,
Through epic altitudes,
And higher latitudes
Falling into sleep,
Drifting into dreams,
Cosmic crashes in my eye,
Cosmic flashes in my brain
Cosmic rays and Wilson clouds,
Clear my consciousness.
Memories of infinity,
Particles of eternity
Starlettes pierce my eyes,
In my brain fire flies.
Periods of light,
Punctuate my night

Cosmic Fireflies
by Story Musgrave (2000)

Japan was the first country to bring poetry officially into its space program. Japanese astronaut Chiaki Mukai (STS-95) began a *tanka* while she was in space on the shuttle. Thousands of Japanese on Earth completed it. *Renshi* is a form developed from traditional Japanese linked verse. Many people contribute to a *renshi*. JAXA sponsored a *renshi* project from 2006, and the resulting poem written by many people on Earth was recorded on DVD and sent to the International Space Station in the Japanese Experimental Module *Kibo* ("Hope").

Space Art and Illustrations

Humans have made pictures of what we see in the night sky for tens of thousands of years. We have also made pictures of what we imagine to be above us but cannot actually see. Fundamental beliefs about humans and not-humans, and about Earth and not-Earth, are illustrated in both such pictures.

The objects above us in the sky seem to move constantly, but often in some kind of a repeating pattern. Knowing what formation the stars take just before the seasons change is important information for any successful agricultural community. This might have been one of the earliest practical uses of astronomical information. Clearly the apparent rising and setting of the Sun and Moon affected humans and Earth. So also might the stars generally, it was often reasoned. Hence, in some cultures – perhaps first in Mesopotamia – there developed a method of predicting the future – of entire societies as well as of individuals in them – based on the movement of the stars, now known as astrology, but an important prelude to modern-day astronomy as well.

Eclipses of the Sun and Moon were especially perplexing since they seemed so out of the ordinary. Eventually careful observers and recorders of heavenly movements noted their regularity and began predicting them with great accuracy.

Partial solar eclipse over the ocean (Graphic courtesy of NASA)

Comets, meteors, and other "falling stars" played a special role in human history, since they exhibit extraordinary astral behavior. In medieval and early modern Europe, and perhaps elsewhere, comets were typically seen as omens of bad things. Comets have been depicted in several surviving visualizations, such as the amazing *Bayeux Tapestry,* which includes Halley's Comet of 1066.

With the invention and spreading use of the telescope, more accurate observations of not-Earth became possible, and their depictions often became more "accurate" as well from the sixteenth and especially seventeenth centuries onward, particularly following the invention of chemical photography in the nineteenth century and electronic imaging in the twentieth century.

Space illustrations, as a self-conscious genre, began at the same time as written space fiction, and many early illustrations were created to accompany and make visual the ideas in written texts. Depictions of space environments, some intending to be factually while others were fantastically presented, flourished throughout the late nineteenth and early twentieth centuries, reaching their heights on the covers and sometimes pages of the pulp fiction magazines. But it was probably the depictions in the large-sized popular picture magazines in the United States, such as *Life, Colliers,* and *Coronet* in the 1950s and early 1960s, that really brought space illustrations to the eye of the public and ignited popular support for space exploration as it became technologically possible for the first time.

Astrophotographs – photographs of the sky, or, more recently of space through telescopes – are often treated as works of art. Indeed, as presented to the general public, astrophotographs, such as those taken via the Hubble Space Telescope, must be viewed *primarily* as works of art since they are framed and color-enhanced for maximum aesthetic effect.

Perhaps the first bit of conscious "art" in space was a small ceramic tile about the size of a postage stamp grandly titled, *Moon Museum.* It was carried on *Apollo 12* (1969). American artists Robert Rauschenberg, Andy Warhol, Claus Oldenberg, John Chamberlain, Forrest Myers and David Novros all contributed to it. In 1971, a small figurine, titled *The Fallen Astronaut* (to commemorate all cosmonauts and astronauts who had died so far) by Belgian Artist Paul Van Hoeydonk was left on the Moon by *Apollo 15* astronauts. A sculpture by Joseph McShane, titled S.P.A.C.E., flew as Payload G38 on the second mission of the Challenger (1984). *Ars Ad Astra: The first Art Exhibition in Earth Orbit* was organized by Arthur Woods and the OURS Foundation in cooperation with the European Space Agency during their EUROMIR'95 mission. This was the most comprehensive exhibit of art in space so far.

As with poetry, there have been a few astronauts and cosmonauts who made illustrations in space. Alexi Leonov, co-commander of the *Apollo-Soyuz* Test Flight of 1975, was one. He also worked closely with Andrei Sokolov, "the dean of Soviet space art." Alan Bean, who flew to the Moon on *Apollo 12,* has described in some detail his aesthetic reactions to the experience, including how he chose the colors, values, shading and other features of his space paintings.

Dancing in Space

If "dance is the only art where we ourselves are the stuff from which it is made," there has arguably been a lot of dancing in space. Few astronauts or cosmonauts have been able to resist the freedom that zero, or substantially reduced, gravity affords. Whether it is Buzz Aldrin bunny-hopping on the Moon or almost everyone turning somersaults in shuttles and spaceships, spacefarers are clearly blessed with Happy Feet.

This is one instance where life has influenced art: Earthbound dancers envy the spacefarers' freedom. Dancers are gravity haters by definition. To dance free of the bonds of gravity would be heaven to them all. But so far no dancers have gone into space. Consequently, more and more dancers have done the next best thing: they have danced momentarily during parabolic airplane flights.

Music of the Spheres

In the West the interrelation between science and music goes back at least to the sixth century BCE philosopher, Pythagoras. His interest in geometry (namely, in the numerical value of lengths, angles, and other properties of lines and spheres), on the one hand, and music (actually, the comparative length of vibrating strings that make differing sounds), on the other, coupled with the assumption that the universe is harmonious, balanced, and perfect, led him to a develop a scheme of geometrical and tonal harmony that pervaded the universe: the "music of the spheres." A century later, Plato elaborated the perspective of Pythagoras. Several hundred years after Plato, the Alexandrian astronomer, Ptolemy, modified these ideas into a form that remained dominant in the West until Johannes Kepler revisited the issue in the sixteenth century, almost 1,500 years later.

The story of music is the same as that of space fiction, poetry, and art. As science and technology changed and deepened human understanding of the universe, so also it provoked and enabled new aesthetic expressions in sound. Franz Joseph Haydn wrote *Il Mondo Della Luna* (The World in the Moon, 1777), an opera buffo. Jacques Offenbach's opera, *Le Voyage dans la Lune* (1875), was inspired by Jules Vernes' story, *De la Terra a la lune* (1872). Though Vernes' story was fictional, both it and Offenbach's opera were meant to reflect and inspire yearnings for real space travel.

As better telescopes and more powerful theories began to tell us newer stories about not-Earth, so also more musicians began to reflect these ideas in their compositions. Frank Fraknoi lists over a 100 pieces of music that make use of serious astronomy. Probably the best-known piece of classical space music – *The Planets* (1916) by Gustav Holst – is often excluded from lists of serious space music because it sought to exhibit the essence of each planet according to how astrology described them, and not according to features understood by current astronomical observation or scientific theory. Paul Hindemith wrote an opera (1947) and then a symphonic suite (1951) called *Die Harmonie der Welt*, based on the life of Kepler.

During the twentieth and early twenty-first centuries, many serious post-classical composers have written space-related music incorporating contemporary as well as traditional instruments and sounds. Typically, these composers use tones, chords, tempi, and intervals not based on the classical ideas of harmony of the eighteenth and nineteenth centuries. Indeed, many of these composers were influenced not only by astronomy and space but also especially by quantum physics, with its emphasis on randomness and complementarity. Their music is "post-Newtonian" in conception and execution.

Some of the pieces above combine sounds recorded from space or spaceships with more or less conventional instruments and modes. But there are a few compositions derived entirely from instruments that recorded various sounds from space.

Popular Space Music

Some popular music has been inspired by a specific space event, beginning perhaps with *Sputnik (Satellite Girl)*, by Jerry Engler and the Four Ekkos (1957). Yuri Gagarin's pioneering flight in the *Vostok* provoked at least two songs in the Soviet Union, *The Constellation of Gagarin* (1961) and *Motherland Knows Her Son is Flying in Orbit* (1961). Similarly, *Happy Blues For John Glenn* (1962) by Sam "Lightning" Hopkins and *The Ballad of John Glenn* (1962) by Roy West commemorated the flight of America's first man in space.

When first Neil Armstrong and then Buzz Aldrin walked on the Moon, the Byrds produced *Armstrong, Aldrin and Collins* (1969) celebrating it, while Jethrol Tull wrote *For Michael Collins, Jeffrey and Me* (1970), commenting on three people who "almost made it," including Collins, who was left on the command module circling the Moon and did not get to walk on it. However it was *Armstrong*, (1969) by John Stewart of the Kingston Trio that caused the greatest stir because his lyrics wondered whether a starving black boy in Chicago or a poor girl in Calcutta knew or cared about the feat since poverty, pollution, war, and hate continued on Earth.

There are two particularly poignant songs about space tragedies. One, *Flying for Me* (1986) by John Denver, was about the *Challenger* explosion. Denver had tried very hard to get on that flight, which was featured as being one of the first to include "ordinary people" among the crew. Jean Michel Jarre's *Last Rendezvous* (1986) was written to be played on a saxophone by astronaut Ron McNair on the *Challenger.*

There has been far too much popular music on space themes to begin to do justice to them here. *Space Cowboy* (1969) by the Steve Miller Band, *Space Oddity* (1969), *Ashes to Ashes* (1980), and *I Took a Trip on a Gemini Spacecraft* (2002), by David Bowie; *Moondance* (1970); *Dark Side of the Moon* (1973) [one of the most popular albums of any kind ever produced] by Pink Floyd, and many, many more. This music is basically "rock and roll" of a specific generation. But the dominant pop music of the present is rap, hip-hop, or indie. These forms, too, have music with space themes. Their roots are partly in rock, but mainly in soul, funk, and to some extent punk – that is to say, primarily in black urban culture.

Filk music is inspired by science fiction and fantasy, and is primarily intended for science fiction fan communities, originating at science fiction meetings in the1930s. Attendees gathering for socializing might sing familiar folk songs. Some participants began setting words relating to science fiction to traditional folk tunes, while new works were also composed and sung. Apparently, when a typo in a magazine article about the practice written by Lee Jacobs referred to "filk music" instead of "folk music," a name for the new genre was born.

However, as said before about other art forms, the only true space music is that which has been composed or performed in not-Earth itself. It seems that the first person to sing in space was the very first person in space: Yuri Gagarin. He is quoted as saying that "When I was going down, I sang the song, 'The Motherland Hears, the Motherland Knows.'" However, apparently the first music to be performed in space was "Jingle Bells," sung and "played" by Walter Schirra and Thomas Stafford on the *Gemini 6* mission, during a radio broadcast to Earth on Christmas Day, 1965.

Astronaut Ron McNair is believed to have been the first person to take a musical instrument – a saxophone – into space and play it, on shuttle flight STS-41B in 1984. As noted above, McNair later died in the *Challenger* explosion in 1986. He had planned to play a work during that flight that had been composed for him by Jean-Michel Jarre. Many other astronauts and cosmonauts have sung and played music in space.

We will see in our chapter on SETI (Search for Extra-Terrestrial Intelligence) that humans have tried to send messages to extraterrestrials in various forms. Often these messages included artistic artifacts (or are consciously art objects in and of themselves).

It is clear from this very superficial overview of space art – written, painted, danced, sung, and performed – that aesthetic expression in many forms is fundamental to the human spirit. It will motivate or "de-motivate" us to leave our cradle. Although some space agencies or companies may or may not encourage or support aesthetic expressions in space, from now on it must be seen as an integral aspect of all space activities, and celebrated – and supported – as such.

References

J.B. Burnell, Astronomy and Poetry, in *Contemporary Poetry and Contemporary Science*, ed. by R. Crawford (Oxford University Press, Oxford, 2006)

J. Clair (ed.), *Cosmos: From Romanticism to the Avant-Garde* (Montreal Museum of Fine Arts, Montreal, 1999)

W. Dingbo, Chinese science fiction, in *Handbook of Chinese Popular Culture*, ed. by W. Dingbo, P. Murphy (Greenwood, Westport, 1994)

A. Fraknoi, The music of the spheres: astronomical sources of musical inspirations. Mercury **6**(3), 15–19 (1977)

A. Fraknoi, More music of the spheres: further astronomical sources of musical inspiration. Mercury **7**(6), 129–132 (1979)

A. Harrison, *Starstruck: Cosmic Visions in Science, Religion and Folklore* (Berghahn Books, New York, 2007)

W.K. Hartmann et al. (eds.), *In the Stream of the Stars: The Soviet/American Space Art Book* (Workman, New York, 1990)

R. Matthew, *Japanese Science Fiction: A View of a Changing Society* (Routledge, London, 1989)

I. Milojevic, S. Inayatullah, Futures dreaming outside and on the margins of the western world. Futures **35**(5), 483–507 (2003)

D. Raitt, ESA and the arts: a programme in the making. Acta Astronaut. **60**, 48–54 (2007)

A. Roberts, *The History of Science Fiction* (Palgrave Macmillan, New York, 2006)

Chapter 3
Dreams, Rockets, Rivalries, and Jobs

In this chapter, we will see that dreams, rockets, rivalries, and jobs combined to lead humanity into the space age – to actual spaceflight and space exploration. Some of the main people and processes that led humans finally to leave their cradle, Earth, and to begin to explore their neighborhood, the inner Solar System, and a bit beyond will be introduced. We will also consider why some of us seem to have chosen to crawl back into our cradle, pull up our blankie, and contentedly suck our thumbs.

As shown in the previous chapter, many cultures have stories about beings from not-Earth visiting Earth, as well as about people from Earth visiting the Moon, Mars, and elsewhere. *Vamana* for India; *Chang-e* for China; *Kaguya Hime* for Japan; and *Icarus* for the West are just a few. For centuries, humans flying anywhere was only a dream, but a strong, compelling, widespread dream.

Leonardo de Vinci (1452–1519) in what is now Italy designed flying ships. Especially gifted – or daring – people sometimes actually tried to fly. It is said (it may not be true) that Wan Hu tried, unsuccessfully, to fly with 47 rockets and 2 kites attached to a chair, during the Ming Dynasty in China (1368–1644). The Wright brothers from the United States may have been the first humans to fly a machine-powered, heavier-than-air craft at Kitty Hawk, North Carolina, in 1903. The Soviet Union was the first country to succeed in actually placing a human in orbit in space – Yuri Gagarin in 1961 – while the first man to land on the Moon was the American, Neil Armstrong, in 1969.

Why did the millennial-long dreams of flying in space finally come true in the middle of the twentieth century? There are three main reasons.

Why Did the Space Age Happen?

First of all, as we saw in Chap. 2, the scientific-industrial revolution of the nineteenth century produced so many new things so very rapidly that many people began to imagine that the future – their personal future and that of humanity generally – might be fundamentally different from, and in fact permanently better than, the

J.A. Dator, *Social Foundations of Human Space Exploration*, SpringerBriefs in Space Development, DOI 10.1007/978-1-4614-3094-0_3, © James A. Dator 2012

present or any time in the past. This produced a new belief in "progress" and "development." This belief became the dogma – not to say the eventual insanity – of "continued economic growth:" the belief and experience of continual social and environmental change that led people to begin to write stories, compose music, paint pictures about the future – the so called science fiction of which space fiction is a part.

The personal and social experience of change led more and more people to expect novelties, to imagine things that had been impossible before, and then to try to create technologies that could make their dreams come true. As we saw, the world's most important and influential space fiction writer, the Frenchman, Jules Verne, and the British writer, H.G. Wells, and others, were specifically mentioned by almost all the early space pioneers as inspirations for their scientific and technological work.

Later, pulp science fiction and space fiction movies, television shows, and popular music inspired generations of people all over the world to dream of space and to support space activities by their nations. The role of space fiction in the creation of space reality is very significant. But space fiction itself could not have arisen and become as popular as it was without the technologies that created the experience of permanent social and environmental change, leading people everywhere to begin to wonder what new things might lie ahead, just over the horizon.

The lesson here is plain: dreams and stories precede and often produce reality. Without prior dreams and stories there may be no new realities. If so, if we wish to create a better world with more successful and adventurous space activities now, we need more and better dreams and stories than we currently have. The old stories don't seem to provide the motivation they used to.

How Did the Space Age Happen?

In addition to dreams, the second thing that led to *Sputnik* and to men on the Moon were developments in science and the engineering of rocketry. Without them, we would still be on Earth, only dreaming.

Key to spaceflight is the reaction principle. This principle was known in ancient Greece. The "pigeon of Archytas," 360 BCE, was a model flying bird powered by jets of steam. Hero of Alexandria described the principles of reaction motion in his treatise, *Pneumatics*, about 2,000 years ago. However, in spite of what the ancient Greeks and others understood in theory and created in prototype, apparently no practical use was made of the principle or of the model flying bird.

Fireworks and firework rockets were developed by the Chinese and their neighbors for fun, spectacular displays, and religious rituals often associated with the time of year crops were planted or reaped. This appears to be the first widespread use of the reaction principle for propulsion. The Chinese developed festival fireworks around A.D. 600 while rocket-powered weapons followed around A.D.1000. This "dual use" of rockets and other technologies first for playful and then for military purposes (or vice versa) has been a continuing feature of space history, as we will see. The first confirmed use of war rockets was in 1232, when China used them against the Mongols. The Mongols, in turn, used rockets against Poland in 1241 and against Baghdad in 1258. In 1249, Muslim armies used rocket-powered projectiles during the Seventh Crusade. By the end of the thirteenth century, rocket weapons were known in Japan, Java, Korea, and India. Knowledge of rocketry spread quickly throughout Asia and into Europe at the same time. In 1379, the Italian word *rochetta* was introduced. This is the earliest use of the word "rocket" in Europe.

Johannes Kepler (1571–1630) and Isaac Newton (1642–1727), so crucial to all aspects of modern science and technology, laid the scientific foundations for rocketry and rocket-propelled space travel. Kepler, a German mathematician, published three fundamental laws of planetary motion in 1609–1619. Isaac Newton, an English mathematician, established the basic laws of force, motion, and gravitation in his *Philosophiae Naturalis Principia Mathematica* (Mathematical Principles of Natural Philosophy) in 1687. Newton's Third Law of Motion was the first scientific definition of the reaction principle.

Later, Hyder Ali (1781) and Tipu Sultan (1792–1799) used advanced war rockets against the colonizing British in India. Indeed, it was the success of the Indian war rockets that inspired Colonel William Congreve to develop a new war rocket for the British in 1804. Congreve's rockets used black powder, had iron casings, and 4.9-m guide stick for stability. They had an average range of about 2,800 m. These rockets were used as signals and as artillery. There is a phrase in the national anthem of the United States about the "rockets' red glare, the bombs bursting in air" that refers to England's use of Congreve's rockets against America during the War of 1812. European and U.S. armies quickly adopted Congreve-style rockets and worked to improve them. The stick-less Hale rocket (1840s) had a greater range and was far more accurate.

Rockets were also applied to civilian use by whalers in the early nineteenth century to fire their whaling harpoons, as well as to launch maritime rescue apparatus. Another example of "dual-use technology." European and U.S. armies lost interest in rocketry in the late 1800s, but their development continued in Russia. The first Russian war rocket was developed by Alexander Zasyadko in 1815 and was used in the Russo-Turkish war of 1825. A rocket works was founded in St. Petersburg in 1820s. Konstantin Konstantinov became director in 1849 and made significant improvements in Russian military rocketry. The appropriately named Charles Golightly obtained a patent for a rocket plane in the UK in 1841, and the Russian Nikolai Kibalchich sketched plans for a solid-fuel rocket space vehicle in 1881. The German Hermann Ganswindt proposed a reaction-powered spacecraft propelled by dynamite.

Rockets were also used to take photographs from a height of about 100 m in1897. These were designed by Alfred Nobel and launched in Sweden. The German Alfred Maul designed camera-carrying rockets for military use. In 1906 he photographed the landscape from a height of 600 m. However, the reconnaissance in rockets was abandoned by the time of the First World War, when aircraft took over. Rockets themselves saw limited use of any kind during World War I. They were employed to lay smokescreens, and some allied forces used rockets to illuminate battlefields at night. The French military introduced rockets invented by Naval Lt. Y.P.G. La Prieur. These were small solid-fueled rockets designed to be fired from bi-planes against tethered observation balloons.

Three "Fathers" of the Space Age

Nonetheless, in spite of many contributors from all over the world, three people stand out as most responsible for making the dreams of spaceflight a reality, and thus may be considered collectively to be the "fathers of spaceflight": the Russian, Konstantin Tsiolkovsky; the American, Robert Goddard; and the German/Romanian, Hermann Oberth. It is important to understand that these first three worked in total ignorance of each other and in general isolation from the rest of society as lonely, eccentric pioneers. They were generally laughed at and ignored – almost always a good sign that they might be onto something important beyond the ken of the common thinking of the time.

Konstantin Tsiolkovsky (1857–1934) might very well be considered the most important single individual in making space flight both a popular dream and a reality. A man of extraordinary vision and imagination as well as of deep knowledge and reflection, he provided a solid scientific basis for space travel. He not only wrote exciting but plausible science fiction, he also made the first designs for multistage rockets and space stations. He designed life support systems and space suits and explained the feasibility of satellites and of solar energy. He mathematically established all the basic laws of space flight, demonstrating that liquid fuel rockets would have the thrust necessary to put a rocket into Earth orbit, or on a journey to other planets. In 1903 he published *The Exploration of Space with Reactive Devices*, the first major work on astronautics. It included the first presentation of the 'rocket equation.'

Most importantly from a space and society point of view, Tsiolkovsky considered space exploration to be an absolute requirement for humanity that would lead to the colonization of the Solar System. He wrote extensively in justification of that belief, arguing it was a moral – almost religious – imperative of humanity to move quickly into space.

It is also important to understand that Tsiolkovsky was as handicapped and marginal a person as one can imagine. He did not come from an influential family. He did not go to good schools and get good grades. He did not live in a population center where he could get from and give inspiration to others. Rather, Tsiolkovsky was the son of a poor wood gatherer who entered Russia illegally from Poland. His mother died when he was young and his father was seldom at home. He was deafened at an early age and never received any formal academic degrees of any kind from any school. He spent most of his life in provincial towns such as Kaluga and Vyatka, far from laboratories and good books. But he left Vyatka at 16 and went to Moscow in order to study in the libraries there.

His brilliance and desire to learn caught the attention of Nikolai Fyodorov, who was the chief cataloguer at the Chertkov Library, and himself one of the greatest geniuses and eccentrics Russia has ever known. Fyodorov enabled Tsiolkovsky to study everything that was being taught in the science and math departments of Moscow University. More importantly, Fyodorov taught Tsiolkovsky Fyodorov's own "philosophy of the common task" – the belief that humans should stop wasting time and intelligence killing one another, or buying and selling trinkets, and learn to transcend the chaos and chance of the cosmos, especially to transcend death. In doing that, humans should move off Earth and go to other planets for new sources of food, energy, materials and living room.

It was this philosophy – this almost religious dream – that inspired Tsiolkovsky to do all the extraordinary pioneering work he did in developing the basis for all scientific, technological, and social aspects of human spaceflight before anyone else did, and well before the technology itself enabled it. In a very important sense, only a marginal but intelligent person such as himself could have dreamed the "impossible" dreams he did, and then had the patience and ability to figure out how the dreams could eventually come true. He did not have parents, teachers, priests, companions, and an entire culture telling him his dreams were stupid, as most of us do. Neither was he told he should become a dentist or a lawyer or an insurance salesman, and stop lying about the house reading and reckoning. He could dream freely and plan brilliantly without crippling ridicule or stifling support.

The second father of spaceflights was Robert H. Goddard (1882–1945), an American physicist and rocket experimenter. In 1919 he published *A Method of Reaching Extreme Altitudes.* It described the first sounding rocket, showing how a rocket might reach the Moon. He was interested in all aspects of rocketry and also in jet propulsion, electronics, atomic physics, and solar heating. Goddard designed and flew the world's first successful liquid fuel rocket on March 16, 1926. Powered by liquid oxygen and petrol, it flew for 2½ s, reaching an altitude of 12.5 m, traveling a distance of 56 m.

Though somewhat less reclusive and handicapped than Tsiolkovsky, Goddard also was a timid and solitary individual. When Goddard's dissertation on spaceflight was published, the front page of *The New York Times* ridiculed Goddard for "failing to know what every school boy knows very well," that rockets won't work in a vacuum with nothing to "push" against. Goddard was so crushed by the ridicule that he left Massachusetts, where he originally lived, and moved with a few loyal friends to Roswell, Arizona where – otherwise alone but with a bit of financial support – he tried to build rockets that never really achieved the goals he wanted. He died relatively young and thoroughly unappreciated for the many important contributions he made to spaceflight.

Hermann Oberth (1894–1989), the third father, was a Romanian-born German rocket experimenter and spaceflight promoter. He also independently developed spaceflight concepts and publicized them in a 1923 book, *Die Rakete zu den Planetenraümen* (The Rocket into Planetary Space). He discussed rockets, space suits, space walking, living in space, and many other aspects of astronautics. Oberth, however, was an important spaceflight popularizer. His books and films greatly stimulated interest in rocketry throughout Europe. Oberth was thus the least isolated of the three, but still basically working alone in developing his ideas and technologies based upon them.

Inspired by Tsiolkovsky, Goddard, and Oberth, an international spaceflight movement began to form in the 1920s and 1930s. Enthusiasts in many countries formed space travel and rocket societies, carrying out theoretical and practical research in rocketry and spaceflight. Some of these groups made very significant contributions to the development of rocketry and astronautics, and nurtured the talents of future spaceflight engineers. But they all were viewed as silly hobbyists and dreamers by most other people at the time and received little attention or financial support.

The Space Age and World War II

One of the most important early spaceflight societies was Verein für Raumschiffart (VfR: Society for Space Ship Travel), formed in 1927. Oberth was a founding member. Many members went on later to develop rocket technology during and after World War II. Most prominent was Wernher von Braun, about whom more will be said later.

In 1921, the USSR established the Gas Dynamics Laboratory (GDL) for rocket research. Ten years later, the Group for Study of Reactive Devices (GIRD) was formed. It designed and launched the USSR's first liquid-fuel rocket. Among GIRD's members was Sergei Korolev, who later would become the leading figure in the Soviet space program.

The Treaty of Versailles, after the World War I, severely limited the kinds and numbers of weapons Germany could have. But as nationalism and hypernationalism began to rise, rapidly inspired by and inspiring Hitler, rockets were seen as a way around the limitations of the treaty. Werhner von Braun and other VfR members were employed by the German army to work on rocket projects. In 1937, the German army created a rocket research center on the Baltic Sea island of Peenemünde. It developed the A-4 (also called V-2: *Vergeltungswaffe* 2, or vengeance weapon). It was a tremendous leap forward in rocket technology. The V-2 established the basic design concepts for rocket motors, fuel systems, guidance, and steering that remain at the heart of today's rockets. Its first flight was on October 3, 1942. More than 3,000 V-2s were launched by the Germans against their enemies during the last stages of the war, causing 2,700 casualties in Britain alone.

Though it was a stunning technological breakthrough, the V-2 was not a successful weapon per se: it could not be guided with accuracy, and its complex technology was unreliable. Its main military value was as a terror weapon. Unlike bombs from airplanes that could be seen and heard well in advance so that evacuation to shelters was possible, V-2's went unnoticed until seconds before they exploded. And the brief screaming sound they made before impact was horrifying.

The V-2 was also enormously important as a harbinger of spaceflight and, more ominously, of the possibilities of intercontinental war. The V-2 was the world's first long-range missile. After the war, it became the prototype for the first ICBMs that played key roles during the Cold War as potential speedy deliverers of world-destroying nuclear weapons. V-2 derived missiles were also developed into the first space launch vehicles – dual use on steroids. It is clear that rockets and spaceflight were minor concerns for most governments and business until it was seen they could be used as mighty weapons. Suddenly millions of dollars were poured into their development for military purposes. Only the Germans were able to bring military rockets online during World War II, but did so too late for their success: the war was basically already decided in Europe before the V-2s could do much good – or bad.

It is worth noting that during World War II, though efforts were made in the United States to develop rockets, the real money and talent there was spent on developing the atomic bomb. The first usable bombs were also developed in the latter stages of the war, and were used by the United States against Japan – devastating the cities and citizens of Hiroshima and Nagasaki. It is further worth noting that many of the people who were instrumental in developing the science and technology necessary to create those bombs for the United States were from Germany, for the most part German Jews who had fled from Germany and Europe to the United States to escape the anti-semitic wrath of Hitler and the Nazis. In contrast, the developers of the German military rockets, like Wehrner von Braun, were "Aryans," not Jews. The German rockets carried conventional explosives. The Germans had no atomic bombs. On the other hand, the Americans used conventional airplanes to deliver atomic bombs to Japan. America had no long-range rockets anywhere near the stage of development that the Germans did.

Imagine what the world would be like now if Germany or America – or both – had had rockets in time to use them to deliver atomic bombs in World War II. Dreams, even bad dreams of hatred and prejudice, drive human activities, and those of us living now can only consider ourselves fortunate that these two mighty tools – nuclear bombs and long range rockets – were not joined in Nazi or American hands before the war was over.

Nonetheless, after World War II, nations rushed to develop their own long-range missiles. Military advantage dominated rocket development, not the desire for spaceflight. As the Cold War developed during the 1940s and 1950s, ICBMs to deliver nuclear weapons became part of the strategic planning of all major nations. They received top priority in both the United States and the USSR, with massive resources being poured into missile development and advanced nuclear bombs.

Indeed, as World War II was ending, the Soviets, moving to Germany from the east, and Americans and its allies moving into Germany from the west, rushed to capture German rocket experts and rocket technology. As it turned out, the Americans and their allies got to them first, and quickly removed most of the scientists, engineers, and rocket technologies to the United States. There was not much left for the Soviets to take back to the USSR, as both countries began investing heavily in rocket technologies for war.

The Cold War

What became the Cold War space race had two leading figures: one in the USSR, one in the United States.

Sergei Korolev (1906–1966) had been a member of GIRD. He became the famous (and for many years anonymous) "chief designer" of the Soviet space program. From the end of World War II, Korolev oversaw development of the USSR's long-range missile program. He was the driving force behind the USSR space program, quietly appropriating military technologies for spaceflight, and finally gaining official support for developing space technologies. His efforts were enormous, carried out almost in secret, not only from the outside world but also within Russia, until his rockets were finally and successfully launched. It is indeed uncertain how the competition for space dominance between the United States and the USSR would have turned out had he not died suddenly, during routine surgery, in 1966, after which the Soviet program floundered.

Wernher von Braun (1912–1977) was the leading figure in the development of the V-2 for Germany. After the war he was brought to the United States along with many of his German co-workers to develop America's missile programs. He did so spectacularly, technologically speaking, enabling the U.S. space program to develop very rapidly from the advanced knowledge and experience he brought with him from Germany. Von Braun was also an extremely effective – indeed, charismatic – popularizer of space travel in the United States, his tall, noble bearing fascinating and attracting more Americans than were repulsed by his Nazi associations.

Scientific upper atmospheric research provided the trigger for the space race. In the mid-1950s, the world scientific community proposed the International Geophysical Year (IGY) to investigate Earth's relationship with the space environment. Both the USSR and the United States declared that they would build scientific satellites as part of their research program.

October 4, 1957, unexpectedly and without previous fanfare, the Soviet Union launched the world's first artificial satellite, *Sputnik-1.* The space age had begun. More than 50 times larger than America's proposed satellite, Vanguard, *Sputnik-1* was a stupendous technological and propaganda victory for the Soviet Union. Startled world reaction to *Sputnik* was spontaneous and concerned. The Russians were suddenly leaping ahead. On November 3, *Sputnik-2* placed the first animal into orbit, the lovable dog, Laika. She proved that it was possible for living things to survive the rigors of launch and the conditions of weightlessness. The Soviet's had no plans to return her to Earth, and her death in space was mourned by millions worldwide.

After the tremendous propaganda victory of *Sputnik-1,* the U.S. government authorized a crash program to develop the Explorer satellite as an alternative to the troubled Vanguard project. Von Braun and his associates, including scientists from the Jet Propulsion Laboratory in California, constructed and launched *Explorer 1,* on January 31, 1958. The satellite was launched from Cape Canaveral using a modified Jupiter C missile.

The United States had expected that it would launch the world's first satellite. When the Soviet Union achieved this distinction instead, it sparked an extreme panic reaction in the Cold War environment of the time. If the USSR could put a satellite into orbit, it was assumed they also had the technology to attack the United States from space. This fear rapidly led to a 'space race' between the two superpowers. Americans responded with frantic determination. The American president, John F. Kennedy, famously declared in May 1961 that before the decade was out, America would land a man on the Moon – and bring him back to Earth. A bold statement! This was something that was technologically impossible when he announced it (especially the "bringing back" part). America had to make good on his boast, or suffer even greater losses of respect and leadership.

The space race was on, with competitive national prestige and security the third factor that led to Sputniks and humans going into space. Political propaganda quickly associated space achievements with ideological superiority: "my political system is better than your political system because my space technology is better!" The United States and USSR vied with each other in a race to achieve status-conferring space 'firsts.' Initially the Russians beat America in almost every aspect in the race to the Moon. Working under a cloak of heavy secrecy and with substantial funding and human resources, Russia achieved first after first, while America's failures and belated successes were widely known and ridiculed.

It seemed that Russia was vastly superior to America in the science, technology, and human abilities necessary for spaceflight. Nonetheless, the Russians failed to attain a huge prize in the race. The Americans were the first to land men on the Moon in July 1969, and to return them safely back to Earth, as Kennedy had promised. America continued to send men to and from the Moon until 1972, when political will and public enthusiasm failed. After Americans reached the Moon, the Soviets gave up the race and turned their attention to other goals, all scientifically and perhaps strategically important, but none as romantic or appealing as the race to the Moon.

To achieve the Moon landing, incremental steps to a permanent human presence in space (set out as far back as Tsiolkovsky) were pushed aside in a headlong rush for geopolitical dominance and propaganda glory. The race to the Moon did not generate the space infrastructure necessary to support a permanent human presence in space, such as space stations and cheap access to space with versatile crew and cargo vehicles. Thus, after the *Apollo* landings, the pace of space technology development slowed markedly and reverted to a more 'normal' type of technological change. Space activities took a step backwards, perhaps providing time to develop the required space infrastructure for a permanent human presence in Earth orbit, and then that had been leapt over earlier but even that opportunity appears to have been squandered.

The space race was over.

Is the Space Age Over?

In 1979, Arthur C. Clarke referred to spaceflight as a "technological mutation that should not have occurred until the twenty-first century." He meant that spaceflight required technologies that had been developed much earlier than should normally be expected. Ordinary technological advances are the products of a gradual process of incremental innovation and improvement. Revolutionary technological change may be brought about by social and political movements operating outside conventional market and scientific processes, and then prove to be unsustainable. Spaceflight was a revolutionary technological change that arrived, driven by geopolitical factors, before the world was ready to exploit its full potential.

It still may not be ready.

The "technological mutation" began with the spaceflight movements of the 1920s and 1930s. The military, political and propaganda requirements of World War II and the Cold War gave added impetus to the development of technologies needed for spaceflight. Massive technical and financial resources devoted to rocket development by Germany during World War II led to the development of the V-2, a major technological leap forward. Cold War tensions after World War II also resulted in enormous technical and financial resources being devoted to the development of long-range missiles. This accelerated the pace of rocket development dramatically so that, by the mid-1950s, ICBMs had achieved the capability to act as space launch vehicles, able to lift scientifically useful payloads.

Many people tell the story of the space race as though it was only between the United States and the USSR. Though less well-known abroad, many other countries were part of the space race, too, and on their own, not necessarily as allies of one nation or the other. The "father of Chinese rocketry" was Qian Xuesen. Qian earned a Ph.D. from the prestigious California Institute of Technology and helped the United States develop jet propulsion and the atomic bomb during World War II. However, during one of the many Cold War "witch hunts" that occurred in the United States, Qian was expelled from the United States in 1955 and returned to China, where he become director of China's rocket research. Fully supported by Chairman Mao, China opened its first missile and rocket research facility in 1956. This is another of the ironies of the space age: political aims and ideological beliefs both drove and handicapped national aims for the development of space and domination of Earth. Initially using Soviet technology, China launched its own R-1 rocket in November 1960, while placing its first satellite in Earth orbit on its own "Long March" rocket on April 24, 1970, broadcasting the song, "The East is Red."

In 1962, with the support of Prime Minister Nehru, Vikram Sarabahi became the father of the Indian space program. With technical support from both the United States and the USSR, Sarabahi was less interested in weapons and races to the Moon, but saw Earth-orbiting satellites as tools for national development that provided communications, remote sensing, weather forecasting, and both educational and popular television broadcasting in support of national goals. The first Indian satellites were launched on Russian rockets in 1975 and 1979.

The French space program, Centre National d' Etudes Spatiales (CNES), was established under President Charles de Gaulle in 1961. The Diamant rocket, launched in 1965, was the first satellite launcher not built by either the United States or the USSR. Other nations developed space programs during this same period as well.

Although there have been spectacular unmanned missions to neighboring planets and into deep space, and while space satellites have revolutionized knowledge about

Earth and vastly improved the quality of life for most people on Earth, humans have yet to return to the Moon or to go anywhere beyond. They seemed content for a while to fly round and round in Earth orbits in little cramped tin cans, calling that a "space program." Now even those are in question.

Jobs for Space on Earth

In spite of many plans and grandiose statements about missions to Mars or back to the Moon, space activities in many countries now appear to be important primarily to provide jobs on Earth – the fourth reason for the space age. In a story that has been repeated thousands of times in every spacefaring country of the world, when President Obama proposed in 2010 canceling NASA's Moon-bound Constellation program, while turning to private companies to launch astronauts into space, the initial reaction was overwhelmingly negative. Republican and Democratic congressmen from Alabama, Florida, and Texas unleashed a barrage of criticism. A Democratic senator from Florida said he would "fight for NASA, and for the thousands of people who stand to lose their jobs," while a Representative who, as a Republican, should have been expected to favor cutting government spending and leaving it up to the private sector, called Obama's plan "a giant leap backwards." A ranking Republican on the Senate appropriations committee, and normally a champion of privatization, declared, "China, India, and Russia will be putting humans in space while we wait on commercial hobbyists to actually back up their grand promises." He said Obama's plans were nothing but "a welfare program for amateur rocket companies." It's all about jobs, baby.

It is very likely that as the private sector increasingly performs services that only the public sector could do before that all of these politicians will become big supporters of the privatization of space. Now that NASA has rolled out its new launcher development program with the less than catchy name of "Space Launch System (SLS)," the criticism has seemed to have abated – particularly since past major rocket suppliers have been given a piece of the pie in the new development.

For now, though, let us thank the dreamers, the rocketeers, and the wars, both hot and cold, that made the space age possible and do everything we can to reinvigorate it with new energy in order that we may live long and prosper.

[Thanks to Kerrie Daughtery, whose ISU core lecture provided the basis for much of this chapter].

References

R. Bonnet, V. Manno, *International Cooperation in Space: The Example of the European Space Agency* (Harvard University Press, Cambridge, MA, 1994)

S.J. Dick (ed.), *NASA's First 50 Years: Historical Perspectives* (NASA SP-2010-4704, Washington, DC, 2010)

J. Harford, *Korolev: How One Man Masterminded the Soviet Drive to Beat America to the Moon* (Wiley, New York, 1997)
J. Johnson-Freese, *The Chinese Space Program: A Mystery Within a Maze* (Krieger, Malabar, 1998)
R.D. Launius, A.K. Johnston, *Atlas of Space Exploration* (HarperCollins, New York, 2009)
J. Logsdon, *John F. Kennedy and the Race to the Moon* (Palgrave Macmillan, New York, 2010)
W. McDougall, *The Heavens and the Earth: A Political History of the Space Age* (Basic Books, New York, 1985)

Chapter 4
Cultural Rationales for Space

This chapter is about the cultural rationale for space activities. Of course all rationales for space activities are cultural. Culture is to humans what water is to fish. Culture is all around us. We may not know it is there until it isn't, such as when we encounter other cultures and discover there are humans who live differently from the way we do, who believe things we don't believe, including, perhaps, that their culture is better than ours!

Culture refers to the way we live, the things we believe, the language we speak, the artifacts we fashion and use. Culture is inescapable. Moreover, we each are part of many cultures, and not just one. There is the culture of our family, our community, our profession, our social class, our ethnicity, our nation, and many more. What may be required in one culture may be forbidden in another. What is considered to be dirty in one culture may be clean in another. A bug may be considered a pest that should be exterminated in one culture, a pet to put in a cage during the summer in another, and when deep-fried with a bit of salt and soy sauce, a tasty snack in another.

We often hear about the scientific rationale for space. We should go into space to increase our knowledge of the universe of which we are a part; to learn more about other planets so we can understand the origin of ours, its distinctive attributes, and its common features with other planets; and to discover many other things that we cannot begin to imagine until we do go into not-Earth and study what we find.

There are also economic rationales for space. Space is a new venue for exploitation and profit. It is a place to establish new colonies with mining and manufacturing abilities so as to feed the insatiable desires of humans for new and better things and life styles so that we can then move on and do the same throughout the solar system and then the cosmos, and any parallel universes we can access.

In which case, there are also policy rationales – laws and policies to govern the interaction of individuals, corporations, and nations as they compete in space as they do on Earth. Policy preferences clearly spring from underlying cultural beliefs and perspectives. The only reason there is a policy on something is that there are cultural beliefs and values that support the policy. What may be considered good policy in one culture might be considered very bad in another.

J.A. Dator, *Social Foundations of Human Space Exploration*, SpringerBriefs in Space Development, DOI 10.1007/978-1-4614-3094-0_4, © James A. Dator 2012

What is now "science" is the current end product of an evolving way of thinking that has existed since antiquity, always in contest with other ways of thinking. That is the case today. Even in so-called "advanced countries," far more people hold views about the world at odds with those of science than adhere to a scientific worldview or process of knowing. Science is a product of some human minds within certain cultural contexts often in opposition to the "common sense" that governs most people's lives.

Economics is also a cultural way of thinking and behaving. There are widely varying cultures within economic thinking as well – some based on ritual gift-giving; some bent on redistributing resources fairly to everyone in the community; some encouraging greed and the accumulation and display of individual wealth in the face of widespread poverty and privation. Economic behavior that is a disgrace in one culture is a source of pride and prestige in another. Economic theories that are held out as "scientific" in one culture are proclaimed mere justifying ideologies in others.

All Rationales Are Cultural Rationales

In short all rationales for space activities are just specific aspects of an overall cultural rationale. The only reason there are space activities, space industries, and space agencies of the kind there are (rather than other kinds) is because of the cultural beliefs about what a government or business space agency should look like. For now, that unfortunately means space must display a serious, solemn, boring persona. It is not possible in the culture of contemporary national space activities for an agency to be named "Yahoo" or "Google" – names that when most people first heard them caused many a giggle and hoot. We may have forgotten that while such names have become commonplace in the private sector, governments remain stodgy, colorless, and perhaps often irrelevant. Cultural rationales for space are fundamental to everything done in space communities.

As we saw earlier, dreams, space fiction, rockets, wars, and jobs all led to humanity finally being able to move, at least for a time, into space. In the early days of the space age, it was the dreamers and fiction writers who dominated. The visionaries and artists who produced paintings, poetry, dance, music, opera, films, and written fiction were the heroes of the day. Space – an alluring fantasy from the ever-expanding future – was an arena of adventure and romance. Then, for a while, amateurs, hobbyists and space enthusiasts ruled. Space was fun, uncertain, and exciting. But unfortunately, since the days of the Cold War, the stern cultures of the military, government, and military-aligned businesses have taken over, and wrung all the fun and fantasy out of the enterprise. Space now is about rockets, wars, and jobs, and not about visions and transcendence. It is serious business for serious people. This "industrial space sector" represents but a tiny fraction of the millions of people who would otherwise be enthusiastic supporters of space activities.

From Nomads to Civilization

For most of human history – for tens of thousands of years – humans lived in small bands and tribes as nomadic hunters and gatherers, living off the land as we roamed. Moving out of Africa, humans eventually came to occupy every spot on Earth, from the most barren to the most lush; from the coldest to the hottest. Although many of us may be stick-in-the-mud homebodies, as a species, we have been migrants, travelers, tourists, eager to know what is over the next hill and beyond the next ocean – eager, and fearful. As we grew in numbers we became increasingly destructive of the natural environment that sustained us as wanderers. Once all new lands were discovered, other humans followed until there were so many people that Earth could no longer sustain our hunting and gathering ways. Our options were to go extinct or transform, culturally and technologically. Some of us learned to settle down, inventing agriculture with the domestication of animals, plants and women. Civilization was born, along with dominating hierarchies of government, schools, religion, and war.

Cultures and the Stars

As we wandered and especially after we settled down, we observed the lights in the darkened skies, and made up stories about what they were and what they meant, which helped tell us who we were and why. For example, what is?

The answer depends on your culture.

My culture tells me it is the Big Dipper, though I suspect that most Americans now are only vaguely aware that once upon a time a dipper was an essential part of every home as a way of dipping water from a bucket or a well. Although the dipper still may be a vital tool for many cultures around the world, water in many homes today doesn't come from a well and bucket. Water comes from a tap, for people of a certain age, and from plastic bottles for others, which horrifies the tap-turners and mystifies the dippers. Culture has changed though we still call those stars the Big Dipper (if we can see them and notice them!). In truth, fewer and fewer of us ever see stars in the night sky at all since light pollution from cities obscures even the brightest of stars and turns the Moon blood red. It seems at times that the only stars we pay attention to now are in movies and on television, not in the heavens.

All cultures in the past noticed those same lights in the sky that some called the dipper, but they may have seen it as, and declared them to be, a plough, a wagon, a bear, a salmon net, a saucepan, the Seven Great Sages, or merely the northern seven stars. Astronomers say they are stars within the constellation Ursa Major, which is Latin (a dead language of a dead culture) for "Big Bear." Each of the individual seven stars themselves have different names in different cultures. In the scientific culture dominant today, they have what are called "Bayer Designations."

Modern astronomy had its origins in astrology, of which there are many versions. Astrology assumes that the movement of the lights in the sky influences events on Earth in some predictable ways. That may seem silly to modern scientific people, until one reflects on how much influence the Sun and the Moon do have on Earth, and that gravity knows no limits. Quantum physics refers to "action at a distance" instantly and everywhere in the universe. So why should not all stars, and especially their movements in relation to each other and Earth, influence events on Earth, some people reckon.

Religion and Not-Earth

If actions in the skies influence actions on Earth, then maybe actions on Earth can influence the heavens? On almost every place on Earth, there are carefully arranged stones that may (or may not) have something to do with the stars, and with the interaction between humans and the cosmos. Those stones may (or may not) have something to do with the origin of religions and the early connection assumed between Earth and humans, and not-Earth and not-humans.

One of the main characteristics of culture – indeed, often the primary distinguishing characteristic between one culture and another – is religion. Religious expressions and institutions often represent humans at their very best, striving to overcome their limitations and living their lives in service to others, as service to God or other higher beings. Some of humanity's most sublime expressions in art, architecture, music, poetry and dance, and some of humanity's most profound thoughts, have been inspired by religious beliefs.

Religion is also a sensitive topic – the cause of many disagreements and even brutal wars between people and cultures. Words are very important to all religions, and a great deal of time is spent and blood is shed in making sure the right words get into the holy texts – especially creeds that are formal statements of beliefs. It is almost impossible even for believers to express the fundamentals of their faith in their own words (rather than reciting exactly from the texts) without committing heresy. So it is with great humility and considerable anxiety that we include here a few words about culture, religion and space. To be clear, the following statements are not, and are not intended to be, officially approved pronouncements, though it is hoped that they fairly represent characteristic positions of the various religions discussed.

Please note the similarities, as well as considerable differences, each of these religious statements make concerning the cosmos and humanity's role in relation to it. Some are quite striking indeed.

According to Hinduism, the universe is made of five elements: water, fire, earth, wind, and space. Space is the element that cannot be seen, felt, smelled, or heard but is all pervading. It is within us and beyond us. There is a mutual interdependence among all things and events that take place in this universe: we are all inseparable parts of the cosmic whole. This cosmic one is intrinsically dynamic – continually alive, moving, changing, dancing! In the fourth century BCE Maharshi Bharadwaj is said to have described three categories of Vimana flying machines: one that flies on Earth from one place to another; one that travels from one planet to another; and one that travels from one universe to another. As we saw, India was among the first nations to use satellites for educational purposes. Two of its earliest satellites were named after famous ancient astronomers: *Aryabhatta,* the first Indian satellite, launched in 1975, and *Bhaskara 1,* launched in 1979.

Ellison Onizuka, who was killed in the *Challenger* explosion, wrote from the perspective of Jodo Shinshu, a form of Japanese Buddhism translated as True Pure Land Buddhism, from about the thirteenth century: "In space," Onizuka wrote, "I saw the Pure Land .... It is the land of 'no boundaries.'" The Pure Land is not a place to go to, like the Christian heaven. It isn't a destination far away or something accessible only after death. According to the founder of Jodo Shinshu, Shiran Shonin, the Pure Land is really just the everyday world around us, properly understood. Thus the "goal" in True Pure Land Buddhism is to become awakened, "to be reborn in the Pure Land," and to achieve the same kind of awakened attitude that Shiran Shonin had, just as Shakyamuni (The Buddha) himself achieved some 2,500 years earlier.

According to Rabbi Nosson Slifkin, from a Jewish position, one of the greatest achievements of the twentieth century was putting a man on the Moon. Now, at the beginning of the twenty-first century, the International Space Station has been completed. Slifkin says that an ancient need that will be assisted by the ISS is the instruction, "Be fruitful and multiply, fill the land and conquer it." (Genesis 1:28). Man is supposed to break new frontiers and expand ever further in God's universe. "He did not create it for nothing: He formed it to be inhabited." (Isaiah 45:18). Voyaging into the heavens should not be seen as intruding into God's domain. There are heavens and there are Heavens. God does not reside in "outer space" – at least no more and no less than he resides here on Earth and everywhere in the cosmos. The Soviets, who scorned God, did not make it to the Moon, while the Americans, who humbly recited the verses attesting to God's creation of Heaven and Earth, landed safely on the Moon in 1969, Slifkin pointed out. "When I look at Your heavens, the work of Your fingers and the moon and the stars, which You have established; what is man, that You are mindful of him?" (Psalms 8:4–5). The ISS is a giant leap for humankind but truly a small step in the greater scale of things, says Rabbi Slifkin.

Br. Alexis Bugnolo has written on the Martian frontier from a Roman Catholic perspective. "The heavens declare the glory of God and the firmament tells forth the works of His hands." (Psalm 19:1). The Martian frontier represents a milestone in the history of human civilization. Ever since our first parents were cast forth from the Garden of Eden, Bro. Bugnolo writes, we have been involved in a struggle for survival against the elements of creation. Unlike the adherents of many other systems of belief, Bro. Bugnolo says, Catholics hold that all things the Creator made are fit subjects of the dominion of man. It is not that the universe has been made hostile by God. Rather, it is that God in His original plan, gave Adam, our forefather, dominion over all creation, so as to glorify the Creator through humankind's efforts to till that garden in which the Lord had planted him from the beginning. The significance of the Martian frontier, hence, lies in this opportunity of renewing the human race in Christ and for Christ, Bro. Bugnolo declares.

Abu Bakar Abdul Majeed explains an Islamic position by reminding us that astronomy, the enabling science of space exploration, flourished in the Muslim world. Science and technology is about inquiry, invention, and innovation in the service of Allah, the Creator. Allah shaped the universe out of nothingness. But his effort was not without purpose. It is up to man to seek out the purpose behind Allah's creations. Nothing is random, erratic or superfluous in its behavior. Everything is governed in a perfect balance and equilibrium. "To Him is due the primal origin of the heavens and the earth. When he decrees matter, He says to it: 'Be,' and it is." (Surah Al-Baqarah: 17). In this modern age of space exploration, it is therefore incumbent upon Muslims to be involved in this venture, says Abu Bakar Abdul Majeed. This is not an arbitrary activity but a necessary one. Allah prescribes in the Quran, "O ye assembly of Jinns and Men! If ye can pass beyond the zones of the heavens and the earth, pass ye! Not without power or authority shall ye be able to pass!" (Surah Al-Rahman:33).

According to Inoue Akio, the path to the Joyous Life originated with Oyasama, whose name is Nakayama Miki. She spent 50 years conveying the teachings of Tenrikyo and providing guidance in order to save all people in the world. In pursuing space development, Inoue says, we need to consider not just scientific and technical matters but also such basic questions as the relation between people and space, between the universe and the evolution of life, and why people exist at all. These questions need to be pursued at least as eagerly as are those concerned with the technology and business of space exploration. Consequently, Tenrikyo (perhaps uniquely among all the religions of the world) has held several space culture conferences with astronauts, cosmonauts, philosophers and theologians discussing these issues.

Ben Finney, Vladimir Lytkin, and others have studied the writings of Konstantin Tsiolkovsky, who we saw earlier envisioned and created in the late nineteenth century much of what was technically necessary for space travel in the twentieth century. Tsiolkovsky was inspired by a worldview called "the philosophy of the common task" that was developed by Nikolai Fedorov, who declared space exploration and settlement to be a necessary next step in human evolution. Fedorov believed that everything in the universe is alive, with humanity achieving the highest consciousness of all life. Thus it is our duty to introduce order and purpose into the chaos and meaningless of nature. Tsiolkovsky expressed this duty compellingly in one of the most iconic statements justifying the exploration of not-Earth: "The planet is the cradle of intelligence, but it is impossible to live in the cradle forever." Humans must grow up, leave their cradle, Earth, and expand throughout the Solar System and out into the cosmos. Tsiolkovksy believed it was his life task to see that humans did so as soon as possible.

Virgiliu Pop has compared the worship accorded to cosmonauts during the Communist days with the importance given spaceflight by the Russian Orthodox Church during the current Russian space program. In many ways, Communism served as a surrogate for religion, so that when it collapsed, traditional religious practices and beliefs quickly took its place and have become integral to the Russian program, Pop says.

Most astronauts and cosmonauts have said that they have a kind of religious experience when they see Earth whole for the first time. Frank White called this the "Overview Effect": "The experience of seeing the Earth from a distance ... and realizing the inherent unity and oneness of everything on the planet. The Effect represents a shift in perception ... from identification with the parts of the Earth to identification with the whole system."

Just as many astronauts have sung, played musical instruments, painted, written poems, and danced in space without permission from ground control, so also have many expressed their religious beliefs, sometimes in contradiction with official protocols and orders. Among the first and most famous (referred to by Rabbi Sflikin, above) was when Frank Borman read from Genesis on Christmas Eve 1968 as *Apollo 8* circled the Moon. Buzz Aldrin performed the Presbyterian service of Holy Communion on *Apollo 11,* and Jeff Hoffman on SST61 (1993) celebrated the Jewish Hanukah with menorah and driedel, but, thankfully, no burning candles. Sultan bin Salnan bin Abdulaziz al-Saud, Sheikh Muszaphar Shukor, and Anoushes Ansari each endeavored to fulfill their daily Islamic prayer obligations while in space. Malaysia has recently developed a manual to guide Muslim worship in space.

At the same time, there also has been at least one astronaut and others who have argued that religion should be kept completely separate from all space activities. They have condemned those who have exhibited their faith in space because it is false, it is unscientific, and/or it is divisive, bringing to space the religious tensions and conflicts they cause on Earth. Nonetheless, religion is too powerful a factor in all cultures to be overlooked, as reasons for or against going into space, and for guiding human behavior in space.

Sex and Space

Rivaling and perhaps exceeding the cultural fervor and excitement cause by religion is sex. So let's consider sex in space:

Actually, though, this is very difficult to do. There is very little reliable data on it. We can be sure that some kind of sex in space has already happened, and a lot more certainly will, but the fun-killing space agencies and industries of the present are predictably puritanical when it comes to discussing sex in space. But once privatization of space activities begins to flourish, you can be certain that sex will be a big money maker in not-Earth, as it is on Earth. There is video proof of sex on aircraft, sex during parachute dives and even bungee jumps. Such precedents strongly suggest that it will be as hard to prevent sexual expression in space as it is to prevent religious expression. Both are fundamental to the human experience. Indeed, sex is clearly foundational to human existence.

Other aspects of the human experience also strongly influence our attitudes and behavior towards not-Earth. If humans had evolved wings, we probably would have been living on the Moon and Mars long ago and perhaps everywhere in the Solar System and beyond. Flying would have been "natural" to us, and so developing technology to enable us to go where our wings could not carry us would surely have been a dream long ago realized. Yet, humans are plodders on Earth, victims of gravity, hesitant even to take to the oceans, much less to the skies. Nonetheless, as Oscar Wilde once said, "we are all lying in the gutter, but some of us are looking at the stars."

Culture and Space Policies

Will humans ever return to the Moon? Will we ever go to Mars? Will humans never return to space at all? Will our own space junk trap us on Earth, even if we want to go again? These are not scientific, or economic, or policy questions. These are cultural issues. We clearly have the technical and scientific knowledge to have been on other planets long ago. Limitations in scientific knowledge and technical knowhow is not what keeps us on Earth. We clearly could find the money for it. Where did the trillions of dollars for bailouts of decades of fiscal folly suddenly come from? Who made the decision to prop up and sustain ever more greed instead of enabling spaceflight? We could make the policies to create and spend the money, and conduct the research and development for exploring not-Earth, if we chose to do so. Money, after all, is simply a human invention. Humans probably spend much more on chewing gum than they do on space, and there clearly is no nutritional or other value to chewing gum whatsoever. We clearly spend far more money on weaponizing space than we do on the peaceful uses of space. We spend substantially more money on watching sports. And we may spend vastly more on pornography than we spend on all of the above put together!

Cultural rationale all!

Now, spending money on chewing gum, watching football, and pornography is largely a personal choice, aided by lots of propaganda and advertising. Spending money on space currently is a matter of national and international priorities. But isn't space just a waste of time, talent, and money? Shouldn't we all be devoting our abilities to solving poverty, ameliorating global warming, preparing for the next pandemic, providing decent housing and pure water for all? Some say that there are useful spinoffs from space, citing the big four of Tang, Teflon, satellite-broadcast Television and Tweets (though arguably neither Tang nor Teflon actually owe their existence to space programs). Wouldn't it be much more cost-effective to apply the money directly to the problems instead of indirectly via space research?

Some say, why not phase out public space activities and turn them all over to the private sector? All governments are deeply in debt because of the fiscal fiascos of the past 30 years. We should not expect governments to fund space exploration very lavishly, if at all, from now on. Of course, countries such as China and India, along with Brazil and others, are determined to become very active in space, and Russia is intent on remaining active also. But the economic, environmental, and energy challenges facing Europe, North America, and Oceania will likely engulf these so-called "BRIC" nations as well. It may be up to the private sector to step forward if we are to have any space programs at all. Space entrepreneurs seem supremely positioned and eager and able to do so. But if nations can't or won't, and if entrepreneurs face financial and other limitations as well, perhaps some religious groups will use their wealth and numbers to keep the momentum going. Some religious organizations have vast amounts of money and command the respect and attention of millions of adherents. Why should they not take the lead?

Some believe it is imperative that humans move into not-Earth, and soon. First of all, two planets are better than one, and more than two are even better. All real science is comparative, and so far we have remained stuck on this one little planet, so we don't know much about anything, and won't, until we grow up and leave the cradle. True, we can learn something by observing the heavens from Earth, or near-Earth. But to really understand the cosmos, we need to get out of the house, cross the street, and begin to explore the neighborhood. Until we do that, humans will remain infants, mere shadows of what we could and should be. Although Earth is our cradle, the cosmos is really our home. We are star children, not just dependents of Mother Earth. We are each made out of the elements formed when the universe itself was created or from stardust that permeated the cosmos when first or second generation stars blew apart. It is time we embrace our essential cosmic nature, and become one with it again.

Earthrise! and Sunset?

About the time humans walked first on the Moon, pictures of something that humans had never seen before appeared: pictures of Earth from the Moon.

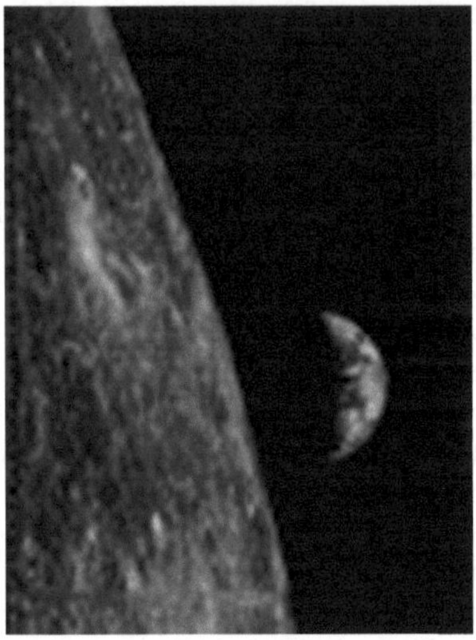

Earthrise!

What a spectacular sight! How awe-inspiring! How proud we should have been. But no! Instead, we gasped and said, "How small Earth is. How lonely it is. How frail it is. How very tiny and lonely and frail we humans are. Don't make me leave home. I want to stay here on Earth forever," many of us said. In the moment of humanity's greatest triumph in its entire history, we lost our nerve and turned inward. It is time for humanity to look outward again as well: Protect and love Earth *while* going boldly elsewhere.

Humans may be at a period of our history – the end of the brief industrial/information era – when we need a grand project to unite us in a peaceful but exciting "common cause," driven by a grand philosophy. What can be grander than space exploration and settlement? While dystopian images of the futures prevail in our fiction and games today, the few positive images of futures in space that do still exist continue to inspire young people everywhere. Whenever children are asked to draw "the future," they inevitably include spaceships and space objects in their images. We should enable their dreams to become reality. Of course, it is not enough only to have appealing images. In order truly to appeal to current and future generations, we must make the reality of space travel not only technically possible, economically feasible, reasonably safe, and philosophically inspiring, we also have to make it fun. Current space reality, in contrast with space dreams, is far too nerdy, serious, and boring. We recognize that space is important to such key functions as telecommunications, remote sensing, meteorology, navigation, and coping with climate change, but this does mean that space must somehow be banned from also being philosophical, artistic and tons of fun?

References

M. Byrnes, *Politics and Space: Image Making by NASA* (Praeger, Westport, 1994)

S.J. Dick, R.D. Launius, *Societal Impact of Spaceflight*. (U.S. Government Printing Office, NASA SP-2007-4801, Washington, DC, 2007)

S. Dick, M. Lupisella (eds.), *Cosmo and Culture*. (US Government Printing Office. NASA SP-2009-4802, Washington, DC, 2009)

P. Harris, *Living and Working in Space: Human Behavior, Culture and Organization* (Ellis Horwood, London, 1993)

F. White, *The Overview Effect. Space Exploration and Human Evolution* (Houghton Mifflin, New York, 1987)

Chapter 5
Humans in Space

One of the biggest challenges for a space enthusiast or a person new to the space literature in comparison with someone who is trained or working in some aspects of space is to distinguish between what we "know" about space from all the movies and TV shows we have seen, games we have played, music we have heard, and (for people of a certain age) books and stories they have read, from what the reality of space exploration actually is. Some of us are so immersed in space as popular culture has constructed it that we greatly exaggerate or overestimate what has already been accomplished by members of the space community versus what has actually been done (much less what is actually underway or seriously planned). When such people discover the truth, they are often shocked, disappointed, and perhaps disillusioned. Space fiction and popular expectations generated by the extraordinary and rapid developments between the late 1950s and the early 1970s are so very far ahead of subsequent space achievements, at least in human spaceflight, that many people have lost all faith in space activities.

Humans in Space

Only a few humans have ever been in space and only very, very recently in the context of human history. As of July 2011, 549 humans have flown in space since the first person, Yuri Gagarin, orbited Earth in April 21, 1961. Of these 549 only 56 were women, with Valentina Tereshkova of the Soviet Union being the first in 1963, though the second, Svetlana Savitskaya, did not fly until 1982, while the first American woman in space was Sally Ride in 1983. Since then the number of women in space has increased, but they are still very much a minority. The longest single human spaceflight was undertaken by the Russian Valeriy Polyakov, who circled Earth for 438 days in 1995 in the *Mir* space station. Most flights have been only of 10–20 days duration, though people tend now to stay on the International Space Station (ISS) for about 6 months, and that may lengthen.

J.A. Dator, *Social Foundations of Human Space Exploration*, SpringerBriefs in Space Development, DOI 10.1007/978-1-4614-3094-0_5, © James A. Dator 2012

So far, human spaceflights have been in two kinds. (1) The early heroic missions preparing for and eventually enabling humans to land on the Moon and quickly come back to Earth (1957 through 1972), and (2) the many practical, necessary, developmental missions when humans have either gone round and round Earth in small containers, such as the U. S. space shuttles, and soon back to Earth, or to orbiting space platforms such as the now defunct *Skylab, Salyut,* and *Mir* and the current International Space Station, where people fly round and round Earth, and eventually come back down again (from 1972 to 2011).

It is unclear whether the flight of the last U. S. space shuttle in July 2011 represents the end of an era and the beginning of another (perhaps missions to the Moon again, Mars, or asteroids), or is a just a phase in the continuation of the practical period with other actors by other means moving forward. Or perhaps it is just another step towards the end of human spaceflight altogether, with either robots taking over from humans, or all attempts at deep spaceflight ceasing altogether, for a while, or forever.

All of the first people in space were either from the Soviet Union or the United States, and some of their political allies. Recently we have seen some new actors – China, most importantly, and a few space tourists. So all we know about actual humans in space comes within the context of very small groups of extremely unusual and atypical people.

Initially, essentially all astronauts and cosmonauts were male test pilots, truly superhuman in every way. They were highly disciplined though exuberant risk-takers who were willing to conform their thought and behavior to test the bounds of a largely unknown and completely new environment within spirit- and bone-crushing capsules. Even when opportunities for experiencing orbital spaceflight were opened to "ordinary" people, they have had to go through long periods of selection, training, and more training that often exceeded the time they actually spent in space (and many, so trained, never had a chance to fly at all). So people who have flown in space so far are not like some random group of ordinary people who get on a scheduled airplane flight from Dayton to Daytona. To the contrary, they are extremely atypical folks in every way.

However, not only is the sample of humans from which we can make generalizations about human spaceflight very small and unrepresentative of humanity, but also the tools of social science that can be used to study and develop valid generalizations about human spaceflight are themselves comparatively new and problematic – scarcely older than the Space Age itself and ethnocentrically limited. In contrast with the natural sciences, which are more nearly globally based, the conclusions of the social sciences have largely come from studies of people in Western countries, as well as Japan and Korea. Even more seriously, most of the space agencies and industries, and especially their sponsors and funders (meaning legislators and/or other political actors), have been extremely suspicious of social science, refusing to fund or even allow properly designed social science experiments or observations to be carried out on space activities. Thus, what we think we do know about human spaceflight comes – with some important but rare exceptions – from anecdotes or inadequately designed and carried out studies about people who have experienced the kind of spaceflight available so far.

Earth Analogs for Not-Earth

Moreover, a very big source of information, or speculation, about spaceflight has come not from actual spaceflight but from analogs – from activities on Earth that are thought to be similar to those in space. So what are the conditions of space for which there are analogs we can study on Earth? Here is a typical list:

- Population: small, homogeneous: all, or mostly, male; mono-cultural; same age-cohort; limited range of occupations.
- Habitat: cramped and crowded; no privacy; visually sterile; constant noise–or utter silence; in constant motion; 24-h "day"; reduced or increased gravity; weightlessness; poor or dangerous air quality; variable air pressure; extreme heat or cold; dust or mold; high humidity; microorganisms or insects.
- External environment: very cold or hot; climate extremes; underwater; underground; toxic atmosphere; solar or other radiation; asteroid bombardment; higher/lower gravity; higher/lower air pressure; all dark/all light; one day longer or shorter than 24 h.
- Isolated from the rest of human society for extended periods.

And here are some places on Earth where humans are now or have been that have some of these characteristics: remote small islands; rural farms; national parks/wilderness; prisons; oil rigs; sailing ships; underwater: divers, submarines, sea labs, aquanauts; underground: miners, cavers; early Pacific island voyagers; early Western explorers; mountains and mountain climbers; desert travelers and settlers; polar explorers and settlers; designed experimental habitat dwellers (e.g., isolation booths and tanks, space simulators, Biosphere 2).

Indeed, are there any things in space without any earthly or human counterpart? *Wait!* Are we sure we have listed the most useful characteristics of space and the most useful Earth analogs? We may have fallen in the trap that the American and Russian space programs have set for us. We have imagined space settlements to be hostile, Spartan, military-like, which they might (or might not) be in the very initial period. But how many people will choose to go to space if it is eternally noisy, crowded, dangerous, and altogether unpleasant?

We must not assume that just because military test pilots were the first persons in space, and because the first spacecraft were so inhospitable, that future human spaceflight, and settlement should, much less must, also be so rugged. Early spacecrafts – perhaps out of necessity, perhaps not – were designed according to criteria that assumed that abnormal humans, who were exhaustively conditioned to accept extremely hostile and inhumane conditions and environments, would function within them. It is curious that even now most actual designs, such that of the ISS, are still extraordinarily cramped and stark, assuming that abnormal humans can and will be made to fit within whatever is designed for them. The reason may be said to be economic necessity, but that seems a lame excuse: probably most people who inhabit space stations drive on Earth a private car that is roomier and more comfortable than their station quarters. No, the reason cannot truly be that we are materially poor.

Could it be instead that we are spiritually and aesthetically impoverished, valuing machines and a kind of "economics" over humans?

The following from *Canada Today* begins to capture the contrast:

> Miners on Little Cornwallis Islands – Since 1981 these Canadians work at the Polaris lead and zinc mines, some 875 miles below the North Pole – one of the northernmost and richest mines in the world. The $150 million investment is expensive because the 200 workers must be protected from loneliness and isolation, as well as the elements. Despite temperatures of 50 below, winds at 70 mph, mining work at 500 ft below the surface, and severe outside travel, there are waiting lists of hundreds to apply for positions from cooks and miners to metallurgists and geologists. This is due, in part, to the very high pay, 2 week vacation flights every 10 weeks on assignment, and luxurious living conditions and food in the residence building which includes basketball courts, pool, saunas, lounges, and jogging tracks.

However, the most useful Earth analogs for space settlements may lie in but one of the places named above, and that is the remote small islands in the South Pacific. These are considered to be as close to paradise as we can imagine. For example, Hawaii is one of the most remote spots on Earth and one of the last to be discovered and settled by humans. Until the invention of the jet airplane, only the rich and leisurely, or poor and outcast, could get there. It was an almost impossible place for most non-Hawaiians to travel to, or to thrive in. Now, mass tourism is the number one, booming industry in Hawaii, and it took a lot of PR to make it so. The Hawaii Tourist Bureau doesn't want anyone to know about the Dark Side of Paradise, and so it paints such a glowing picture that tens of thousands of people each year endure long, uncomfortable flights and dump bucketsful of money into the economy simply because "Hawaii" has come to mean "Paradise" to most of the rest of the world, whatever it may mean for those who actually live there.

That is to say, Earth analogs we should be looking for should be those that express the very best aspirations for freedom and community we humans have ever imagined or dreamed of – and created. Not-Earth should be portrayed as "Hawaii," not as hostile wasteland where all must live as prisoners. But that is not the kind of analogs we have studied. Since the immediate future of human spaceflight may well be one of spirit-crushing habitats in challenging and unfamiliar environments, all Earth analogs to space have had those characteristics, and that is a big PR mistake. It might be acceptable for now, but if we really want "ordinary" people to go to Mars and elsewhere, we need to stop calling space "hostile" or "an extreme environment," and do what the Hawaii Tourist Bureau does – call it "a paradise," with great stupefying drinks, nubile women, and rugged, but understanding, men. Human perception of an experience (called the "definition of the situation") is in some ways more important than the "situation" itself. Some things, if approached with the expectation of being positive, are experienced positively, while the same situation, if defined negatively, will be experienced negatively. Moreover people's initial attitudes towards an environment typically change with prolonged exposure to it.

Social Science About Human Behavior in Space

Social science studies based mostly on limited social psychology and/or small group theories have been used to observe and understand human behavior and thoughts while on human spaceflights and in analogs on Earth. Even so, very little social science-specific research has actually been carried out on actual space flights. This is so not only because funders don't like it but also because it has been judged to be dangerous for any one person's future in space if their behavior is revealed and they are found somehow wanting (as all people are) on some or several tests. Candidates for spaceflight and astronauts in space resent, and indeed are fearful, of all tests that might reveal weaknesses on their part. This is true of the medical tests they must necessarily endure but even more so of any social science tests since they have faith in the former, but not the latter. Consequently, human thoughts and behavior are the least studied aspect of space studies (in terms of funding, number of studies, number of researchers, replications, etc.). Conventional physical/geological and biological/environmental natural science aspects of space are the best studied, followed by issues of designing, constructing, operating and maintaining spacecraft. When humans are studied, it is primarily in biological/medical terms with a bit of psychobiological or psychiatric or maybe psychological matters thrown in. There have been very few properly designed, funded, executed, analyzed and published social psychological studies. The few that do exist are those called "human factors" that seek to measure and improve human work productivity, and those related to organizing and managing a space activity in order to further the political or commercial interests of the sending/funding organization.

There appears never to have been a political (or bio-political) study except in terms of the national and international law involved in deciding to do or continue a space activity. That is to say, there have been studies about the legality of a space activity, but very few concerning the politics surrounding or within the activity. Many excellent biological and physiological experiments useful for human spaceflight have been performed on the ISS, but apparently no specifically designed social science ones, in spite of the fact that the multinational, multicultural, and to some extent multidisciplinary nature of the crew makes it a perfect laboratory for all kinds of useful social science research.

Specifically, NASA has something called the Human Research Program. Its six programs are titled the International Space Station Medical Project, Space Radiation, Human Health Countermeasures, Exploration Medical Capability, Behavioral Health and Performance, and Space Human Factors and Habitability. Researchers in the latter unit study "how the design of the spacesuit, spacecraft, or habitat affects crewmembers. They also develop new equipment, procedures, and technologies designed to make the space environment more comfortable and livable. Recent design innovations include isolated sleep quarters that offer privacy and reduced noise, as well as personal space for writing emails, relaxing with a movie or a book, or engaging in other leisure activities."

There is also a National Space Biomedical Research Institute (NSBRI) consortium of universities and others focusing on "health concerns facing astronauts on long missions" so as to "benefit health care on Earth." The mission of its research on Human Factors and Performance is "to reduce performance errors and mitigate habitability, environmental and behavioral factors that pose significant risks to mission success." Research is "examining ways to improve sleep and scheduling of work shifts as well as how specific types of lighting in the craft and habitat can improve alertness and performance. Other projects address improving the interactions between automated and manual control of a spacecraft and how factors in the environment, such as dust, can impact crew health."

These are all necessary and proper research projects, but not much that would interest most social scientists or humanists, or that will shed light on topics of human spaceflight of interest to them, and probably equally vital to the success of a mission.

It should be pointed out that at the very beginning, NASA did a great deal of psychiatric, psychological, and social psychological research primarily to determine who should be admitted to the astronaut corps, and to evaluate astronaut performance, but apparently all the data was destroyed, in part to protect privacy and in part not to provide possibly unwelcome answers to questions about who had and who didn't have "The Right Stuff." NASA should ramp up such research once again if it does get involved in long-duration human spaceflights, but it is by no means clear that it will.

It is also the case that America's antipathy towards social science and the humanities is not shared worldwide. Even though the social sciences within the Soviet Union were quite primitive and restricted compared to those in the West during the Cold War, impressive research in some social science areas was done for the Soviet space program, perhaps because the Soviets were planning for and involved in multi-member, multi-cultural crews for somewhat longer-duration flights well before the Americans were. Social science projects, and even humanities programs, have been part of space projects in Europe and Japan, though no national program has done (or at least published) the kind of research into class, gender, ethnicity, sexuality, and politics that would likely be done by a sociologist or political scientist. For example, as suggested earlier, we know nothing, because our space agencies will not allow us to know anything for sure, about sex in space. Sexuality is about as fundamental a part of being human as is imaginable, but all space agencies pretend it does not exist. Nonetheless, satellites (the major money-making and personnel-employing activity of all space agencies and businesses) have made pornography cheap and abundant on TV and the Internet, while one viable business model of private spaceflight might involve sex tourism or space "love boats." All this is considered taboo for space agencies to talk about officially.

We said earlier that cultural differences influence human interactions. What may be acceptable in one society may be forbidden in another. Different preferences in food preparation and eating, body odors, social distance and many more things sometimes result in conflict on Earth. For example, Kung, Mehinaku, Arabs, Javanese, Malays, and Japanese, in that order, seem to prefer much more human closeness and contact than do most Americans. Americans maintain a physical distance between themselves and others that seems "standoffish" to many, while the Arab's desire for close contact (and smells) alarms most Americans, who are constantly backing away as friendly Arabs advance. What Americans may consider to be necessary "privacy" is deep pathology to others. Attitudes towards drinking behavior and drunkenness; sexual comments and touching; how to be sick and respond to others who are sick; and many more behaviors are very different from one culture to another (including not only national culture, but also gendered, occupational, religious and other subcultures), but none of these important cultural differences have been specifically studied in space, though they have very frequently been commented on anecdotally and hence unscientifically.

Technology influences behavior in important (and barely studied) ways. In the old days, when movies were shown to people in Antarctic winter-over settlements via 16 mm projectors, everyone watched them together as a shared communal experience. With videotapes, and then the Internet, the experience is now often private. The loss of socialization might be important.

Social Science Findings About Humans in Space

In sum, and with thanks to the work of the people listed in the final paragraph of this chapter, here is what the data we have suggests so far about human spaceflight:

As has already been indicated, the earliest astronauts and cosmonauts during the first, heroic phase of space travel were exceptional people with exceptional characteristics. They were expected to suffer extraordinary pain and suffering, mental as well as physical, and not only endure it but also to thrive in it, often alone or in the company of similar physically and psychically strong humans. This was not only expected of them, but they expected this of themselves. As one shrewd observer commented, even during "the Space Shuttle era NASA seems to have been intent on presenting the astronaut corps, males and females alike, to the public as a homogenized group of clean-cut, all-American 'good guys' (leading some to call them 'the Stepford astronauts')." As it turned out, no one "cracked." There were no failures during the early human space programs because of human weakness or mental health. They all clearly had the "right stuff" and did as they were programmed to do.

Nonetheless, with the end of the space race to the Moon and the beginning of the era of orbiting platforms with larger crew sizes, a different type of spacefarer was needed, one who was both a strong leader and also understanding, sympathetic and nurturing. Or at least some crews with some mix of cultures and genders appeared to be better with a person who was a decisive leader in the initial phase of a project, but who became a companion, or perhaps father figure, later on, enabling others to exercise leadership and initiative.

Many of the analog studies show that there are typically four phases in any long-duration project: an initial period when everyone is excited, energized, focused, disciplined, learning new things; a second period when things become routine, perhaps boring; a third period when ennui, discontent, and perhaps rebellion sets in; and the final period when euphoria returns as preparations for return begin, ending almost always (but not always) in exultation and triumph when the project is over, and everyone returns victoriously "home."

Different people – or at least people playing different roles – are needed during each of the phases. There needs to be a clown or jester who is good at relieving tensions with humor without causing or provoking ridicule or harm; a counselor who is good at listening, sympathizing without judging (often a woman, if there is one in the crew, though women often report this role gets very burdensome if they are the only female, since most men may want to talk freely and be consoled but few are willing to listen in exchange); and someone good at leading decisively when needed, but sharing leadership whenever possible, if appropriate. A contested generalization says that men tend to be task-oriented while women group (process)-oriented. Crews on long-duration missions without people playing such roles can break down into warring camps, sometimes actually sabotaging the project. This apparently has never happened on actual spaceflights but is not uncommon on analogs and so might occur in spaceflights of the future.

What Happens Without Good Social Science?

Nonetheless, there have been many examples of crew discontent, rebellion against ground control, and the like even among those with the "right stuff." Here is one example, told in some detail since it illustrates many of the points made so far, including the uncertain, anecdotal aspect of much of our "knowledge" about human spaceflight: The American astronaut, Norman Thagard, trained extensively in Star City, Russia, for a mission with a Russian crew on the *Mir* space station. He could speak Russian and was quite familiar with Russian culture. He was interviewed on his birthday, and his 111th day in space, about his experiences on board *Mir.* Thagard was reported as saying, "My impression is that psychological aspects probably loom the largest. There don't seem to be any big problems physiologically." Thagard said that there are "problems of cultural isolation among crews of mixed nationality." He said that he would sometimes go for 72 h without speaking English, and that he was very tired of Russian-style food. (*Honolulu Star-Bulletin*, July 3, 1995, p. A-7).

This statement produced a storm of protest. An article (*Honolulu Advertiser*, July 7, 1995, P A-1) titled, "Russian hosts 'shocked' by astronaut's whining," reported that the Russian newspaper, *Izvestia*, said that "the cosmonauts regret Norman Thagard's scandalous interviews. The common assessment was that he had been whining." The newspaper said that "Russian space officials saw that the 115-day flight had become difficult for Thagard during the televised interviews with the crew, when the astronaut appeared passive and reluctant to speak. 'However, it was hardly reasonable for Mission Control to switch into English, 'it said.' And it was impossible to set a separate table for the American,' the *Izvestia* story" concluded.

Thagard was also quoted as saying "I think anybody can do 3 or 4 months. Six months and longer is a different matter entirely. We need to address some things for the folks who plan to be up here 6 months and longer." *(Space News*, July 10–16, 1995, p. 8) NASA Administrator Daniel Goldin promised quick action. "I think this is going to turn out to be one of the major findings of this mission," he said. "If we expect to send people on missions of 2 or 3 years, we darn well better deal with the psychological aspects in addition to the physiological ones. This hasn't been our tendency in the past" he concluded. Tom Sullivan, mission scientist for the Atlantis flight, said Thagard's observations were similar to those of researchers spending winters in isolation in Antarctica. "It's just culture shock," Sullivan said. "When you go to any foreign country, whether its Russia, Japan or whatever, you're going to be out of sorts. That's a well documented phenomenon. It's just isolation. 'I think if we pay attention and provide some of the amenities, the communication with home, the camaraderie of crew mates you can relate to well, all of those things will go a long way toward providing an … environment that will be easier for the crew.' Sullivan said."

It should also be noted that an editorial in *Space News*, July 24–30, 1995, p. 24, stated that the *Izvestia* reports were "overblown." "As one of the Russian cosmonauts put it last week, 'Certain journalists like to use an element of sensationalism. Sometimes rumors are taken as reality'" Nonetheless, the editorial found the discussion useful: "Such topics were a far cry from the rote astronaut-speak of a group of highly skilled professionals who attend NASA's own version of charm school. More typical catch phrases of shuttle crews are those that stress what an honor it is to be part of a terrific team. Who would argue with that? But who can learn anything from it? After scores of spaceflights, the public rightly yawns at such prepackaged happy chatter. As space enthusiasts worldwide ponder ways to revitalize interest in space endeavors, they would do well to take notice of what got the public's attention this time. The human interest aspects of manned space flight cannot be ignored. NASA must stop worrying about institutional perceptions. Astronauts should not worry about how their comments will play in the worldwide media. Let the public see and hear first-person accounts of human spaceflight without the glossy editing. Indeed, the naked truth may generate far more support for space missions than the standard 'Golly, it was great' statements that form the bulk of public flight debriefings" the *Space News* editorial concluded.

An article in *Spaceflight*, September 1995, reiterated much of the above, but added additional important information: Thagard, and his two companions on *Mir* performed 29 separate science experiments on themselves in seven research areas. Only one was labeled as involving "behavior and performance." "However," the *Spaceflight* article noted, "the science programme was far from easily conducted as evidenced by the comments and behaviour of the Mir-18 crew who were described as 'testy.' Thagard complained at one stage that the treadmill tests were too hard for a crew completing an almost 4 month long mission. Thagard refused to wear the halter monitor more than once because of the discomfort of shaving his chest repeatedly for the electrodes to be stuck on. Mir-18 commander Dezhurov actually refused to wear the NASA LBNP and even complained to the Kaliningrad [Mission Control Center] about the experiments, whilst Strekalov balked at certain procedures involved in the experiment. NASA's science team re-planned their schedule constantly to accommodate the changes. There was much discussion about the attitude of the three men during these tests and a NASA spokesperson even ventured that the men were volunteers and could pull out of the tests if they wished. It was noted, however, that Thagard, despite his protestations, had followed the protocols of the scientific experiments." (p. 313. See also an interview with Thagard in *Countdown*, July/August 1995, pp. 50–55).

All of this is very, very familiar. Both anecdotal stories and attempts at social science research show that the entire history of manned spaceflight is replete with episodes exactly like these, but consider what Tom Sullivan, mission scientist for the Atlantis' flight, was quoted as saying about culture shock and isolation. First of all, he says that Thagard was experiencing "culture shock." How does he know this? One does not get the impression from the *Countdown* interview with Thagard that he was culture shocked. It appears his command of the Russian language was good, and he had spent some time in Star City before he joined the cosmonauts in *Mir.* But then Sullivan said, "Its just isolation." But that is not the same as "culture shock" at all. Moreover, this was not Thagard's first mission in space. He should have been prepared for isolation. Then Sullivan apparently said: "I think if we pay attention and provide some of the amenities, the communication with home, the camaraderie of crewmates you can relate to well, all of those things will go a long way toward providing an … environment that will be easier for the crew …"

Here we have a bundle of more problems. Is it a question of "the amenities"? The literature is full of references that state that the interiors of Russian spacecraft are always functional but, from an American point of view, ugly and crowded, with all of the wires and tubes exposed. American spacecraft are neat and clean – antiseptic, in fact. If it is a question of "amenities," whose aesthetic is going to be satisfied? And whose sense of safety and easy repairability?

Concerning Sullivan's other point, there was for a long time a very lively debate in the literature about whether it is good to have people in space in regular communication with their families at home or not (especially when there are crises – even deaths – at home). Now, of course, video and email communication between astronauts and their families and friends is routine, if restricted. And yes, if there is anything the literature seems to confirm, it is that it is the interaction of the crew (a question of sociology) rather than the mental toughness of the individuals in the crew (a question of psychology or psychiatry) that is the more important. And yet the only thing that got some attention in past crew selection was tests of mental condition, rather than attention to group dynamics, or the equal importance, possible primacy, of group interaction over individual mental soundness.

Social Scientists and Space Research

The point in deconstructing Sullivan's apparent statement is not to focus on him (indeed, the statements attributed to him may be untrue or out of context) but to make it clear that there is always some kind of "social science" or "social analysis" going on. If it does not involve trained social scientists, doing the best they can to be accurate and useful, then it will be engineers, managers, media flacks, spin doctors, or politicians who will use "common sense" (also called "folk sociology" or even "crackpot realism") to analyze and "fix" whatever social problems might be anticipated or actually happen.

Consider it this way: Would you want a respected social scientist with no engineering training of any kind (but who had played lots of space war games and talked with astronauts) to design a rocket ship? If not, then you also should not want a rocket scientist implicitly to "design" or explain social interactions within the rocket ship by default. Nor should social science research be entirely a matter of "human factors engineering," which tends (to overstate it a bit) to try to fit humans into the hardware rather than designing the hardware to facilitate behavior humans prefer. This is part of what good social science research should strive to do. Another part of good social science research should be to "problematize" the entire space venture, which is even less welcome, but nonetheless absolutely necessary from a space and society point of view. There have been almost no critical analyzes of space programs or projects by social scientists, and this is a shame. Good social science is essential.

If, as we move from Earth's orbit to the Moon, Mars, asteroids and beyond, the mission goals, composition of crews, internal and external environments, role of robots, artificial intelligence and other autonomous entities, and many other things will present us with challenges and opportunities unlike those we have experienced so far on Earth or not-Earth. We will need to prepare for novel situations that will be encountered by individuals, by small groups, and by organizations. Foremost among the latter will be opportunities for new forms and processes of governance that have seldom been imagined much less developed and used on Earth. We will consider that in a later chapter.

As we saw, one of the most frequently mentioned human consequences of spaceflight so far is the so-called "Overview Effect" – that transformational change of values, attitudes and often behavior that many astronauts and cosmonauts have said they experience or have exhibited as a consequence of seeing Earth for the first time from afar, whole, beautiful, precious, and without any political or other artificial borders and boundaries. Certainly the astronauts who walked on the Moon (even more so those who had to stay utterly isolated and alone in a hovering craft while others walked the Moon) were awe-struck by Earthrise and the glimpse of the tiny, frail, solitary Little Blue Marble that is Earth. Moreover, as we saw in the chapter on space art, many spacefarers have felt compelled to dance, sing, paint, write poems and stories, and to pray. Perhaps as more humans experience spaceflight they, too, might have such spiritual and intellectual transformations.

On the other hand, it is possible that passengers on trips to Mars, an asteroid, or other more distant objects will feel something quite the opposite as Earth recedes to a mere dot, or disappears entirely, before their new destination comes dramatically into view. It is likely that humans circling and then landing on Mars will experience something far beyond any "overview" effect as they realize humanity has finally gotten out of the cradle and begun to explore the neighborhood of the inner Solar System, and eventually even beyond.

References

N. Kanas (ed.), *Psychology and Culture During Long-Duration Space Missions* (International Academy of Astronautics, Paris, 2009)

B.H. Kevles, *Almost Heaven: The Story of Women in Space* (MIT Press, Cambridge, 2006)

D. Lempert, Living in space: cultural and social dynamics, opportunities, and challenges in permanent space habitats. Astropolitics **9**(1), 84–111 (2011)

D. Manzey, Human missions to Mars: new psychological challenges and research issues. Acta Astronaut. **55**, 781–790 (2004)

P. Santy, *Choosing the Right Stuff: The Psychological Selection of Astronauts and Cosmonauts* (Praeger, Westport, 1994)

D. Valoch (ed.), *Psychology of Space Exploration*. (NASA SP-2011-4411, Washington, DC, 2011)

Chapter 6
Humans, Robots, Artificial Intelligence, and Autonomous Entities in Space

Humans and physical technologies have interacted intimately throughout all of human existence. Humans as such have never existed "without technology." Indeed, humans become "human" and change the meaning of being human in large measure by interacting with each other and their environment through technologies. For much of human history, our technologies, though necessary tools for human activities, were largely or completely dumb – dependent on humans to function. Over time, we have also made smarter and smarter technologies, able to do more on their own, if properly designed, built, and maintained.

Many religions have objects that are believed to have special powers beyond what may seem obvious on the surface. Garments touched by saints have saintly properties. If properly blessed by people who can bless them properly, bread and wine become the actual body and blood of Jesus – not just a symbol but real. Stone images of fertility goddesses, marble sculptures, crucifixes, portraits all have sometimes been more than their mere physical selves. When the objects have been able to move, seemingly by themselves, their importance in human lives increased. In the late European medieval period, clocks and then mechanized mannequins served as harbingers of a myriad of increasingly mobile and automated devices from the late eighteenth century onward. At the same time, humans became increasingly dependent on mechanical, chemical–mechanical, or electro–mechanical devices to tell us what was happening with our machines, ourselves, and the world around us.

Instrumentation is vital for many aspects of modern life, especially airplane flight and eventually networked airline systems. By the time of the first spaceflights – i.e., *Sputnik* and beyond – developments in electronics enabled sophisticated automated devices without which human spaceflight would not have been possible. Humans and increasingly powerful robots have always been part of space programs.

A great deal has been written recently about the increasing use of robots in space, either interacting with humans or operating on their own. Most of it is very practical, focusing on the limitations of current robots and artificial intelligence (AI),

J.A. Dator, *Social Foundations of Human Space Exploration*, SpringerBriefs in Space Development, DOI 10.1007/978-1-4614-3094-0_6, © James A. Dator 2012

which often necessitate close, or final, human supervision and control, but which also require careful thought, design, testing, and preflight programming and subsequent deprogramming. We will discuss some of these ideas in this chapter, but we also will use futures studies to imagine later in the book some more speculative, exciting, but nonetheless feasible, ideas about the future of space. These explorations include human futures while working with and collaborating with "smart robots," AI, and even the implications of post-humans in space.

"We Shape Our Tools and Thereafter Our Tools Shape Us"

It is clear that humans in their current biological state are fit only for Earth, and that in fact extraordinary measures of many kinds need to be made to enable them to operate anywhere other than on the thin surface of Earth. Humans need help even to go into the oceans that cover most of Earth. Humans so far have extended their very limited "natural" abilities by prosthetic means – by developing technologies. As Marshall McLuhan said long ago, "The wheel is an extension of the foot; the book is an extension of the eye; clothing, an extension of the skin; and electric circuitry, an extension of the central nervous system." This has been one of humanity's most distinctive features, creating technologies to do what our biology, psychology, and sociology cannot "naturally" do. Humans grew from the frail "hairless apes" of a hundred thousand years ago into the dominators of the planet today primarily by technological and not by biological evolution.

As far as space is concerned, we cannot fly "naturally," so instead of growing wings, we developed airplanes. Even if we had evolved biological wings, like any bird we cannot survive beyond the lower atmosphere, and so we eventually developed spacecraft and spacesuits (i.e., personalized spacecraft). Once in space, humans cannot survive in open settlements on the Moon, the asteroids, at Lagrangian points, Mars, or – as far as we can tell – anywhere in the Solar System and beyond (*Star Trek* episodes to the contrary notwithstanding) without expensive and extensive protective covering or by "terraforming" the environments – making them as Earthlike as possible.

Thus, in a way, humans have for eons been intimately interacting with robots every day in many ways. Humans have long been "cyborgs," (cybernetic organisms) with a biological base and a prosthetic superstructure. It began with clothing and shelter, developed into carriages and boats, augmented by eyeglasses and tooth fillings, and on to the present with hip-replacements and cosmetic surgery to where we now have become almost totally dependent for the very basics of our daily lives on increasingly complex, intelligent, and anticipatory technologies – smart cars, smart mobile communicators, smart clothing, houses and environments – technologies that make decisions for us that once only humans could make, looking out for us, anticipating us, protecting us, surveilling us, and advising us about even the most intimate aspects of our lives.

Right Stuff or Right Robots?

It is odd then that some in the space field find it strange that as we move into space we must do so with robots and artificial intelligence (AI), or that at some point in the future – perhaps not very far off – it may make far more sense to have autonomous entities (AE) in space instead of humans as currently biologically constrained. On the other hand, perhaps we will choose instead, or also, biologically to modify old humans so they can thrive in environments for which they have not yet "naturally" evolved.

As we established in Chap. 3, the decision by President Kennedy to land a man on the Moon and return him safely to Earth was not made as it might more "naturally" have been made. A "normal" science and technology-driven space program would have followed a long, slow, scientifically solid process. It might have begun with the careful observation and mapping of the Moon first from Earth via increasingly powerful techniques; then by a series of closer observations from Moon-circling unmanned spacecraft; then landing robotic rovers on the Moon that carefully surveyed all of the surface (and below) while performing a series of experiments, perhaps then robotically preparing a base (or series of bases) where humans with their AIs/AEs would eventually arrive, set up shop, engage in more exploration and scientific experiments; finally evolving into a permanent and expanding settlement to which "ordinary people" (and AIs/AEs) from Earth might migrate for ordinary or extraordinary purposes; and so on to exploring and eventually building settlements of humans, transhumans, AIs/AEs on Lagrangian points, perhaps asteroids, perhaps directly to Mars.

But no. Absolutely to the contrary, the Moon landing was not a process of such a "normal" scientific, technological, industrial, and human evolution but rather a bold/rash/stupid political decision made entirely for the purposes of international ideological rivalry alone, with almost no proper scientific-technological-human preparation or justification. Very importantly, while technology did exist or was created that got the assigned job done on time, well, and at enormous expense, other technologies that could have made the process less costly and more sustainable – including more advanced robots, AIs/AEs, cyborgs, transhumans, and post-humans among other things – did not exist at the time. If currently looming resource, ecological, and political limitations do not make it impossible, these technologies should eventually emerge, transforming Earth and enabling the steady exploration of the Solar System and cosmos.

However, that is not what happened. Instead we went to the Moon without much in the way of robots and AIs but rather with heavy dependence on brute technologies and brave, intelligent, and utterly abnormal humans. When the ideological purpose was achieved with the American landing on the Moon and the space race won and lost, long-range space programs everywhere came to an abrupt end. In the early 1970s, both the Soviets and the United States ended their premature programs of long-range human spaceflight that many expected would continue (settlements on Mars by 1980!!). After a period of uncertainty very similar to the present, both the

Soviets and the United States (soon joined by Europe, Japan, and Canada) eventually settled on an uneasy balance between human space programs of Earth-orbiting spacecraft on the one hand and robotic space programs of Earth-orbit-based space observation platforms and Solar-System exploring and deep-space voyaging vehicles on the other.

Space: Just an Ordinary Job

Subsequent decades of work on *Skylab, Salyut, Mir*, and the ISS mentioned in Chap. 5 routinized certain space activities while atrophying, technologically and psychologically, the once impressive "space muscles" that were built-up during the space race era. There have been many important technological developments, as well as improvements in human/robot interaction during this time (for example, the robotic "Canadarm" on the ISS working together with spacewalking astronauts, which, on its own, not only helped complete the construction of the ISS itself but also enabled the spectacular Hubble Telescope repair mission). However, from the point of view of most people not directly involved in the space industry, space seems to have become just a regular job, done off-Earth perhaps and so requiring a big, expensive and cranky bus with skilled bus drivers and passengers who go back and forth to work at a science lab that has rather spectacular views where routine science and technology-testing is performed. The space program has become just another high tech job done in a fairly remote place, but a place that is not psychologically more remote than it is, say, for someone who lives in Hawaii and has to fly to Europe and back frequently as part of her job. The generally safe and predictable nature of space station activities is, perhaps, one reason for relative lack of popular support for or interest in space now. It is just routine "going to work" for some people in the space industry. Nothing more. This conclusion is reinforced by the fact that when the routine is broken, and disaster strikes (as in the catastrophic explosions of the *Challenger* and *Discovery* space shuttles), America's interest in spaceflight spikes sharply up, only to settle back down to its very low level when normality and predictability returns. It is hard to be excited about most of an astronaut's day job.

With the end of the space shuttle period, the future of American participation in human spaceflight activities is more uncertain than ever before. Like *Apollo* before it, the shuttle program was canceled before a new American program was determined, much less underway. Some in the U. S. space community now expect that the private sector will step forward to supply the service the shuttle did to keep the ISS fully operating – now that it is finally finished! – well beyond its currently planned ending date. In the meantime, highly reliable Russian rockets will be used, while other nations – for example, China or ESA – might choose to offer taxi service to and from the ISS, guaranteeing a human presence in at least Earth orbit for a while longer.

Since China was not allowed to become a partner in the ISS, as it wished, it has begun construction of a space station of its own. China and India seem to have their eyes on the Moon, and Mars as well, and apparently intend to generally follow the "step by step" method the Americans and Russians more or less used. In the meantime, the private space sector might focus primarily on suborbital space tourism while the United States develops the capacity to return to the Moon, visit one or more asteroids, or prepare for Mars. In those ways, too, America could continue to participate in human spaceflight activities.

Humans and Robots Together

From the earliest days of spaceflight, humans and automated systems have co-existed uneasily. Yuri Gagarin had no real control over his craft. He basically was just along for the ride while machines did the flying. The same was true of the first American manned flights as well, though some had the ability to alter automatic commands in case of an emergency. Fortunately, no early astronaut or cosmonaut felt the need to do so.

The *Apollo* spacecraft were initially designed to carry humans to the Moon and back almost entirely automatically. Human intervention was only to occur in an emergency – as it did. The proud test pilots NASA engaged demanded at least some measure of "flying" control. Similarly, the shuttle was designed to land automatically, but the humans who were to be on board demanded that they be able to guide the shuttle to a landing themselves, necessitating hundreds of hours of training for a skill that would be used only a few times at best and need not be exercised at all.

Most robotic spacecraft so far have been used for scientific purposes such as taking photographs or occasionally obtaining and analyzing soil samples, for example, the Soviet *Lunokhod 1,* the first robotic lunar rover in 1970, *Viking* landers on the Moon in 1976, and the Mars rovers Sojourner (1997), Spirit and Opportunity since 2004, along with the Japanese *Hanabusa* that sampled the asteroid Itokawa in 2010. On the other hand, the intrepid *Voyager 1* and *2* spacecraft were designed to make a photographic Grand Tour when the outer planets of the Solar System were aligned as they rarely are. Each of the *Voyagers* got gravity assists from the planets, enabling the two spacecraft to fly by and photograph Jupiter, Saturn, Uranus, and Neptune, and their assembled moons and surface features. The *Voyagers* then continued sailing far past the planets of the Solar System into deep space. In the highly unlikely event that they might encounter intelligent beings on the way, they each were outfitted with primitive, problematic, and probably completely unintelligible but nonetheless well-intended messages informing the lucky extraterrestrial (ET) what they are, where they came from, and what life on Earth was like, according to the message-writers at the time the two *Voyagers* were launched.

Unmanned supply vehicles routinely arrive at the ISS. They often dock entirely autonomously, though sometimes humans intervene with manual control (and docking has not always been100% successful either way). Sometimes the human guidance comes from Mission Control on Earth, sometimes from humans on the ISS. This is an example of the area of research called human supervisory control

(HSC) where humans do not continuously monitor or directly control a robotic operation but only periodically, or as needed. As human/robotic interaction increases, more research should be devoted to designing and refining this interface for all future spaceflights. Some future robots are expected to be exploring scientists, but some also might be designed and built to perform many tasks associated with construction – specific site-selection, surveying and ground preparation, and erecting structures suitable for human habitation on those sites. Other robots might perhaps repair or upgrade existing, or build new, spacecraft.

Almost all current space policies and research seem to consider robots to be merely devices that have controllers, sensors, actuators, communication capabilities, and a power supply. They are designed to do jobs that are too tedious, dirty, delicate, intricate, or dangerous for humans to do. AI is also called upon to do tasks that it takes humans too long to do or that humans can't do at all because they are too complicated or require undivided attention beyond the abilities of most or all humans. In addition, specialized human knowledge has also been offloaded onto what were once called "expert systems" that enable intelligent robots to make routine judgments, perform surgery (even telesurgery), and engage in a myriad of other crafts and skills that once only humans could do.

Robot Ethics

As many people know, the term "robot" comes from a Czech word for "forced labor" and was apparently first used in a play by Karel Capek titled "R. U. R.," or "Rossum's Universal Robots," in 1921. Sometime later, the science fiction writer Isaac Asimov used the word "robotics" and in 1942 wrote a robot story "Runaround," which contained his now-famous "Three Laws of Robotics", namely:

- A robot may not injure a human, or, through inaction, allow a human being to come to harm.
- A robot must obey orders given by human beings except where such orders would conflict with the First Law.
- A robot must protect its own existence as long as such protection does not conflict with the First or Second Law.

Recently there have been several serious discussions about ethical relations between humans and robots, especially as robots become more and more humanlike in behavior and more responsive to humans. Some of these discussions have started with Asimov's laws as the ethical basis, but soon found these were inadequate from two points of view. One is that they don't apply to what are often the most advanced of all robots now – those intended for military service. By their very design, they are intended to violate Asimov's first law. They are supposed to capture or kill humans. On the other hand, if they fully follow the third law, they are useless on the battlefield, and indeed there were reports a few years ago of military robots that refused to fight because they realized that they were putting their own continued existence in danger, and so they retreated as the enemy advanced. They had to be reprogrammed, just as any human in boot camp is, to overcome their fear of death or injury, to fight without becoming foolhardy, and to follow the orders of their commanders without question. It is a tricky distinction difficult for both humans and intelligent robots to master. As military operations move more and more towards war via robots, with fewer and fewer humans, it is hoped that wars will finally be fought according to the international law of war and rules of engagement, since robots can be programmed to obey those laws more fully than can humans, who sometimes lose their temper and do barbarous, outlawed things to other humans.

Asimov's laws may also be inadequate for another reason: as robots/AI and the rest become increasingly intelligent, emotional, and hence lifelike, they may begin to demand rights similar to those accorded other intelligent living beings, such as humans. One of the earliest and most frequently cited reports on the topic, written for the Hawaii State Judiciary by the Hawaii Research Center for Futures Studies, is entitled "The rights of robots: Technology, law and culture in the twenty-first century." It was written by two futures researchers, Philip McNally and Sohail Inayatullah, while they were interning at the Department of Planning in the Hawaii State Judiciary in the mid 1980s. They traced the evolution of rights of humans from the days of Athens when only certain men had rights, while women and slaves had none, to (in America) the slow and tortured recognition that human rights should be granted not only to certain ethnic, cultural or "racial" groups or males, but to all racial groups and women. They argued that the extension of rights to artificial intelligence was a probable and logical next step following the attainment of universal human rights, and then the rights of children, animals, and even of landscapes and trees.

The issue of the rights of robots and AI was further developed recently by the American jurist Frank Sudia, among others. Sudia uses the term "artilects" to describe "artificial intellects" who may soon demand, and receive, their rights. Robot rights may differ from the rights accorded humans, since artilects, while intelligent, may have needs and wants different from those of humans. The rights of robots and ethical relations between humans and AI/robots has also been discussed seriously by governmental committees in the UK and Korea. This may seem like a silly concern, but, like many ideas that are first rejected and ridiculed, may eventually prove fruitful, useful, and someday "obvious." Ethical relations between humans and artilects may indeed be vital for the success of future space missions. Long-range spaceflight missions that are planned and carried out without sufficient consideration of the ethical and emotional relationships between robots and humans could be doomed to failure, or worse.

One illustration of a continuum of human/computer interactions that might also be extended to clarify robotic, AI, and human relations is this:

1. The computer offers no assistance: humans must take all decision and actions.
2. The computer offers a complete set of decision/action alternatives, or
3. Narrows the selection down to a few, or
4. Suggests one alternative, and
5. Executes that suggestion if the human approves, or

6. Allows the human a restricted time to veto before automatic execution, or
7. Executes automatically, then necessarily informs humans, and
8. Informs the human only if asked or
9. Informs the human only if it, the computer, decides to
10. The computer decides everything and acts autonomously, ignoring the human. (Sheridan and Verplank cited in Liang Sim)

At the present time, rovers on Mars, for example, are extremely slow, cautious, and restricted. Scores of humans monitor their every extremely deliberate move because it takes about 40 min for a query to reach a rover on Mars from Earth and for the rover to respond, while a response by a rover to an Earthly command may also take a long time, as all systems are checked and perhaps double-checked and conditions reported back to Earth before the commanded maneuver is carried out by the robot. Telerobotics currently is thus slow and very costly.

As robots get more intelligent and autonomous, some imagine that humans might come to play the role similar to that an overseer on a plantation. The robots would be like plantation workers, cheerfully picking grapes, bananas, cotton, sugar cane, or pineapples as instructed, only occasionally needing to be corrected, repaired or "disciplined" by the human overseer, riding comfortably in his or her electronic horse in orbit above. That is to say, some suggest that, rather than both landing on the surface of Mars (or Venus), humans and robots would travel to those planets together while only the robots would actually descend to the surface. There they would perform pre-assigned tasks more or less autonomously under close human observation and ultimate control, but with considerable autonomy and discretion.

Although increasingly dependent on automated technologies with increasing intelligence for more and more aspects of their daily lives, a few humans most of the time, and most humans some of the time, resent the powerlessness that comes when machines perform and often outperform what only humans once could do. When allowed to, humans thus may intervene in automated processes when they don't need to, perhaps without serious negative consequences. But sometimes the consequences can be disastrous, such as the collision of two airplanes because one of the pilots did what ground control told him to do rather than what his onboard computer told him to do, which turned out was correct.

Computers are improving in their ability to make accurate split-second decisions and actions, but humans do not seem to be improving in their abilities. Future humans might very well become improved cyborgs, better equipped because of their symbiotic relation to increasingly advanced artificial intelligence and prosthetics to cope, decide, and act. The first people to land on the Moon relied heavily on laminated paper maps and mission control instructions to walk on the Moon. Future Moon or Mars walkers – or orbiters – can be expected to have much better prosthetics that will enable them to see, hear, feel, perhaps taste and smell, and to walk, roll, or fly on extraterrestrial environments far better than humans can now or could previously.

Almost everyone assumes that missions of these kinds will continue increasingly to involve robots and emerging artificial intelligences and autonomous entities working as extensions of and in concert with humans in space. But some voices are urging that these missions, at least initially, be carried out entirely by robots and AIs. On the one hand, it is much cheaper and safer to send robots instead of humans into space, and on the other, robots, AI, and AE are becoming smarter and smarter, more and more mobile, and hardier and hardier. Indeed a minority in the space community (but more in the futures community) argue that AI/AE will rapidly replace human intelligence in the near future, a process that Susantha Goonatilake first identified as "Merged Evolution," and more recently Ray Kurzweil and his supporters call "The Singularity" when accelerating technologies in so many fields merge and transform Earth and not-Earth.

Beyond the Uncanny Valley

Many of us give our automobiles, motorbikes, bicycles, or computers affectionate names. We talk to them, pat them, show them that we care and appreciate them, holler at them when they misbehave. Many humans develop an even closer interrelation with certain animals such as dogs, cats, and horses. We often love our dog more than our children, or our horse more than our spouse, because of their undying loyalty and devotion. It is no wonder that we develop special affection for objects that seem to respond to and actually care for us. And they need not look human at all to evoke strong ties of emotional bonds from us.

Once upon a time, AI was touted as valuable because it could make decisions rationally without being distracted by human emotions. That still is the case. But at the same time, artilects have been developed that appear to sense how we are feeling and respond to us sympathetically, as though they care and share our pain. Some of them are made in the shape of a human, though most of them are still in the "uncanny valley" of creepiness – they look and act almost human, but not yet fully so. In some ways, those AIs are more repulsive than attractive, compared to a box that simply says nice comforting and ego-boosting things to us as we engage with it. Robonaut 2, currently on the ISS, is still caught in the uncanny valley, but more successfully anthropomorphic robots may soon proliferate. The human-machine interface should also quickly and vastly improve, becoming more and more human-friendly. Instead of typing in instructions or pulling levers and pushing buttons, humans should become able to interact with AI by speaking naturally, merely looking, gesturing, or via direct brain control – all capabilities under very promising development now.

At the same time, each generation of new humans will perhaps be born into environments with progressively more ubiquitous, intuitive, responsive, interactive, and intelligent technologies that will be completely "natural" to them. We might expect some current human resistance or clumsiness to disappear (though perhaps new phobias or objections will also arise) as younger generations accept them without question. At the present time, there is the notion of the "digital native" – those people who grew up with multi-functional mobile devices and social networking technologies – compared with those who grew up earlier with television, and even earlier with books, radio and newspapers. Most members of the older generations find it increasingly difficult to master what comes naturally to the younger ones, and when they do, still "speak with an accent" like any immigrant does who struggles to master a language that their native-born children master without effort.

Also, more and more humans are choosing or would be willing to incorporate technology into their bodies – electronic chips, skin-implanted sensors, synthetic body parts and organs, an extra heart just in case since one is often not enough, radiation-repellant skin, perhaps a protective tortoise-like shell into which to retreat from solar flares, wings, perhaps, tails – whatever modification someone might want might well be possible. Goonatilake's *Merged Evolution* and Kurzweil's *Singularity* might be closer than many in the space field think, and can be made even closer if they wish and work to make it so. The old distinctions between organic and inorganic, between living and nonliving, between material and mental and spiritual seem to be fading away. Recently, Lee Cronin spoke of inorganic life, and no one is giggling.

References

C. Breazeal, *Designing Social Robots* (MIT Press, Cambridge, MA, 2002)
A.R. Chapman, M.S. Frankel (eds.), *Designing Our Descendants: The Promises and Perils of Genetic Modifications* (Johns Hopkins University Press, Baltimore, 2003)
S. Goonatilake, *Merged Evolution: Long-Term Implications of Biotechnology & Information Technology* (Gordon & Breach, London, 1999)
P. Harris, *Towards Human Emergence* (HRD Press, New York, 2011)
R. Kurzweil, *The Singularity Is Near: When Humans Transcend Biology* (Viking, New York, 2005)
R. Launius, H. McCurdy, *Robots in Space: Technology, Evolution, and Interplanetary Travel* (Johns Hopkins University Press, Baltimore, 2008)
M. McLuhan, *The Medium Is the Massage* (Random House, New York, 1967)
P. McNally, S. Inayatullah, The Rights of Robots: Technology, Law and Culture in the 21st Century. Futures **20**, 119–136 (1988)
H. Moravec, Rise of the robots – The future of artificial intelligence. *Scientific American* (2009), http://www.scientificamerican.com/article.cfm?id=rise-of-the-robots. Accessed 10 Aug 2011
L. Sim et al., Past, present and future implications of human supervisory control in space missions. Acta Astronaut. **62**, 648–655 (2008)
F. Sudia et al., Essays on artilects. J. Futures Stud. **6**, 43–108 (2001)

Chapter 7
Governance for Space Settlements

> "I can say only this!" Arkady said, staring at her bug-eyed. "We have come to Mars for good. We are going to make not only our homes and our food, but also our water and the very air we breathe – all on a planet that has none of these things. We can do this because we have technology to manipulate matter right down to the molecular level. This is an extraordinary ability, think of it! And yet some of us here can accept transforming the entire physical reality of this planet, without doing a single thing to change our selves, or the way we live. To be twenty-first-century scientists on Mars, in fact, but at the same time living within nineteenth-century social systems, based on seventeenth-century ideologies. It's absurd, it's crazy, it's – it's – " he seized his head in his hands, tugged at his hair, roared "It's unscientific! And so I say that among all the many things we transform on Mars, ourselves and our social reality should be among them. We must terraform not only Mars, but ourselves." [Kim Stanley Robinson, Red Mars, New York: Bantam Books, 1993, p. 89]

Yes, it is crazy – and "unscientific" – indeed. But it is certainly true. Few people in the official space community are thinking about governance systems for space, and when they do, present Earth law prevails over future space governance. We focus overwhelming and blindly on forms and processes invented in the eighteenth century and still in control today.

Probably the most out-of-date aspects of the everyday world we all live in now are our systems of governance. Families, businesses, religions, education, communication, transportation, beliefs and ideas – everything – have all changed in form and substance over the last 200 years. But not governments. They are, as the fictional person, Arkady, says, "nineteenth-century social systems based on seventeenth-century ideologies," based on pre-industrial technologies and cosmologies. And they have a firm control over our minds and actions! At the current time, the government of the United States is locked tight in stalemate as factions seriously argue over which side is most closely following the words and intentions of the Founding Fathers who wrote the U.S. Constitution in 1789 – as though that matters in 2011, when almost everything else about the United States and the world is utterly different. As great an invention as that document was 200 plus years ago – and it was a great invention indeed – the question is its relevance today. Many would argue that given the shift in technologies, social systems and even our cosmologies, a 200-year-old

J.A. Dator, *Social Foundations of Human Space Exploration*, SpringerBriefs in Space Development, DOI 10.1007/978-1-4614-3094-0_7, © James A. Dator 2012

document is not appropriate to twenty-first century needs and opportunities. And yet, all subsequent attempts at "constituting" a democratic polity have basically followed the forms and philosophies of that long-ago era.

Society Is an Invention

Society and all its parts are inventions. The institutions of society are not "givens" in the way the natural environment once was. Everything in society was created by humans and needs to be re-created by us as times and situations change. It is necessary for us to become social inventors, especially of governance systems. And yet nowhere in our formal educational system are we taught to become a social inventor. At best we are taught to be "social scientists" – to study what is, not to imagine and create what could be. Like an architect that is trained to design and construct a functioning building or community, so we also must learn to think in terms of what can or should be, not about what will be or what is likely to be. The future of governances cannot be predicted. It can and should be envisioned and invented by each generation so that each can live their lives to the fullest, individually and collectively.

Humans have had many chances to update its once-great eighteenth-century political systems – most notably at the end of the Second World War, when new constitutions were written for Germany, Italy, and Japan and for scores of former colonized countries in Africa and Asia. When communism collapsed at the end of the 1990s, the questions about appropriate forms of governance abounded. Currently the European Union is struggling to find – so far unsuccessfully – a suitable political form. Each of these pivotal political events provided wonderful opportunities to re-think forms of governance. But that did not happen at all. Instead, old forms of governance were refurbished and re-implanted, or foreign (mainly U. S.) forms were transplanted, more or less intact. The unsatisfactory nature of the results are plainly evident.

New Governance for Space?

So what about space? What should be the governance systems for space settlements? There will likely be phases or stages of exploration and settlement. First will probably come a few intrepid explorers under strict ground control. There will be some, but limited, opportunities for new forms of governance for them. Then there might be military/economic exploiters (including the first hardy tourists), perhaps also with some scientists and their support staff. Here new design options are greater. But finally, when ordinary humans come seeking extraordinary human experiences should be the best chances for envisioning and creating new governance systems.

One of the big opportunities of space – indeed one of its major reasons for space settlements – should be to imagine, invent, create, and re-create new forms of everything.

Not-Earth is different from Earth in so many ways, it is indeed folly to think about transplanting anything from Earth into not-Earth without first carefully thinking of alternatives. We have the opportunity – necessity – of a new, do-it-ourselves Genesis. As Arkady said, it is an unprecedented opportunity indeed.

Although we are not yet even in phase one of our movement into not-Earth, almost all (of the little) thinking about space governance has long seemed content to export intact or modify only slightly existing systems of governance, usually relying entirely on one's own political experience, which is to say either the presidential system of the United States or else the more common parliamentary systems of the world. If the only possible governance system is a choice between parliamentary and presidential systems, then parliamentary systems must win, hands down. Fred Riggs and others have shown very convincingly that presidential systems inevitably lead to a stalemate, which soon leads to military dictatorships, while parliamentary systems are significantly less likely to do so.

Although an argument can probably be made that military-like and Earth-centered command structures are necessary, or at least acceptable, during the initial stage of planetary exploration (although good arguments can be made against that as well), clearly some other forms of governance should, and probably will, emerge, either by design, evolution, or revolution once larger, permanent settlements arise. While the very earliest settlers might be content to follow the rules as established by Ground Control, the universal example of colonial settlements on Earth – exacerbated by the vast distances and differences between Earth and not-Earth – strongly suggests that spacekind will separate from Earthkind, and that even the earliest space settlers may soon try to establish their own forms of governance, compatible with their own wishes and desires. Given the novel challenges and opportunities of space, they will not imitate any presently existing form of government. So, what forms of governance might they consider?

We're Made for Small Groups

They might begin by recollecting that good anthropological evidence suggests that for the overwhelming majority of humanity's stay on Earth, humans have lived in very small bands or tribes of from 20 to 300 people, or in villages of the same size (and only rarely more than 1,000 people or so). These groups were genetically, linguistically, and thus cosmologically homogenous. They were nomadic, peaceful, egalitarian, bountiful. Most people lived in what has been called "subsistence affluence" without "working." Humans evolved to live in such a world and not the new world we live in now. Given a chance, many humans seem to want to return to it, and not-Earth will, at least initially, provide that chance.

It is important to remember that the world in which many of us live now is itself very, very recent. Even as late as the eighteenth century, "large cities" often had only 5,000–20,000 people in them. Philadelphia, where the Founding Fathers gathered in 1787 to write the U. S. Constitution, had less that 30,000 people at that time, and yet the Fathers complained bitterly of the filth and congestion of what we now might consider to be a very pleasant little hamlet. Mammoth population growth and urbanization is a very recent phenomenon indeed. At the beginning of the nineteenth century one of the largest cities in the world was Tokyo, with slightly over one million people. London, with under one million, was the largest in the West at that time. Now, Tokyo has over 35 million people while all of the ten largest cities in the world are located in Asia, Africa, or South America, except for New York City, with about 20 million. For the first time, more of humanity lives in urban areas than in rural ones. True, some civilizations did produce impressively large cities, and some sustained them for a while, but they remain the exception. "Civilization" – life in cities – is only a few thousand years old – a blink of the eye for the lifetime of *Homo sapiens* and absolutely nothing since the origin of life, or of Earth, or of the cosmos. Thus it may not be too much to say that humans are evolved for small, face-to-face groups where decisions are "democratically" made.

So what happened? Why are we no longer living in peaceful and bountiful hunting and gathering societies? Roughly 10,000 years ago, some human populations exceeded the carrying-capacity of their environment. In a way, this was nothing new. The record is clear that earliest humans routinely overpopulated and over-exploited their environment. But they usually could move on to other environments, which they then over-exploited. When they could move no longer they either died out locally, or else invented technologies that allowed them to exploit as resources elements of their environments that were not seen as, or needed as, resources previously.

States, Kings, Empires

So humans began to settle down, developing horticulture and eventually agriculture. Land ownership became important. Heredity rulers emerged. Eventually civilization arose – vast rural areas based on agriculture supporting an urban hub with its centralized, hierarchical rulers; formalized bureaucracy for a formal state; formal military with organized warfare and authorized killing; and priests who enforced official beliefs and organized religion. Eventually empires arose, controlling vast areas with large populations by means of organized violence and killing, all of which was made possible by the invention of writing.

As human populations grew and expanded, this form of governance spread over the world with a few monarchs, nobles, warriors and clergy ruling over many feudal peasants until the emergence of the nation-state and the international system several hundred years ago, first in Europe and then spreading, primarily by conquest, to the rest of the world. The nation-state system itself rose largely in response to the new and growing industrial, merchant, bourgeois and middle classes against the interests of the old land-owning monarchs, nobles, and clergy.

New Classes, New Governance

The new classes wanted a voice in their governance, which the monarchs, nobles and clergy resisted. Moreover, how was it possible to allow so many people to participate in governance? Their numbers were too large. They were too spread out geographically also. They were mainly focused on local issues, and not those of the nation-state per se. Very importantly, there were few roads for transportation (people had to rely on walking or travel on horseback), and no means of communication save speech and writing.

The answer to the many obstacles towards expanded participation in governance was a great political invention that is still with us – representative government. The newly important citizens could elect a local person they knew to "re-present" them. Their representative would then go to some central place to meet with other local representatives and make decisions that then bound everyone, using the principle of "majority rule," even though consensus was the preferred method but too time-consuming to be effective in their fast-paced world. Representation and majority rule were great social inventions of the time, based on cutting-edge technologies and cosmologies of the time – primarily the printing press and Newton's mechanics applied to society as well as to nature. In spite of the rise of the ideas and influence of Darwin, Freud, Heisenberg, Foucault, and Daly who have revolutionized our thinking and overturned most old practices in many aspects of our lives, Newton's old ideas still dominate law and politics, creating the mismatch between governance and other social institutions of today.

Old Cosmologies, Old Technologies

So here we are in the twentieth century with seventeenth, eighteenth, and nineteenth-century ideas and practices of so-called "democratic" governance touted worldwide as the best we can do. They are not. They were arguably the best we could do then. And they should be respected for that. We should also carefully learn from them, especially the process by which new forms of government were created in the late eighteenth century. But now we can and should try to do better.

Direct Democracy?

One frequently discussed alternative is direct democracy, now most frequently as electronic direct democracy. If self-government without representatives was once possible, why isn't it possible again, if people wish it. But would self-government (direct democracy) work for large groups? It might, via forms such as "teledemocracy," described below. There have been many experiments using television and interactive media in order to try to achieve "electronic direct democracy." More recently, the so-called "social media" have given millions of people worldwide the experience of self-government and self-control that futurists have always held out for electronic communication technologies. To be sure there are many technical design problems that need to be worked out, and many have been discussed, but these problems are surmountable if we truly want to design an effective electronic democracy. The main obstacle is that those in power now wish to keep controlling others, but people's self-governing experiences with the social media may eventually be too powerful to resist.

It is often said that many ordinary people are not interested in politics. It is true that many people are not interested in the pseudo-participatory forms of governance that are characteristic of current governance forms. Any intelligent person must know that merely "voting" does not do much to effect policy in a positive and lasting way. However, people could become "interested" in political participation if they knew their participation counted (and that their non-participation counted even more). Very few people are apathetic about everything. Almost everyone has areas of life that excite and animate them and in which they are deeply involved. These are always activities in which their participation and non-participation matters. Politics could be one such area for almost everyone if participation were made easy, effective, and fun, and if non-participation were made costly to their interests.

Are ordinary people informed enough to have direct democracy work? Can we trust people to vote intelligently? Won't they respond only to emotion and demagoguery? The answer here is similar. If people can participate easily and effectively, and if they learn that they will be directly and quickly affected by the consequences of their decisions and non-decisions, then most will become serious and informed before they make a decision. Few polities take seriously the political education of their citizens, preferring docile sheep to informed decision-makers. Learning fairly and fully about important social issues and how they can be discussed frankly and interestingly within the political community must become a central task of education – and of mass media. Indeed, the entertainment and communication media are probably more effective modes of education than is formal schooling presently. Re-thinking schooling in our media-rich environment is a necessary part of rethinking governance.

How can political participation be made more effective and inviting, without becoming trivial? There have been many experiments that suggest that there are ways if we choose to use them.

We can start by asking what are the most popular forms of presentation in the world today, especially on television? They seem to be comedy in various forms, game shows, staged "reality" shows, soap operas, and sports. Social media also demonstrate that self-education and peer-education is highly effective, however dubious the content may be. We also well know what is not popular – what most people will avoid if they have a choice: boring "talking heads." It is possible and desirable to use the tricks of the most sophisticated advertisements, media productions, and electronic games to help people understand complicated issues accurately and fairly, and to let them participate in exciting and satisfying ways.

Ted Becker and Christa Slaton pioneered and developed a technique called "televote" that combines informing people fairly and interestingly while polling them on their attitudes, and then actually having them vote their preferences. It is based on the fact that most people form their opinions, not in isolation, but after talking them over with family and friends. But most polling techniques call people blindly and ask their opinion on things they may never have thought of until the pollster contacts them. Televote gives people a chance to look at interesting, balanced material, talk about it with their peers, and then give the pollster an answer.

Of course television has been surpassed by and merged with electronic games and the Internet into new forms of interactive individual, group, as well as truly mass (national/global) media. Current experiences online with cyberdemocracy, smart mobs, and the like have encouraged many people to wax optimistically eloquent over the future of worldwide direct political participation on the Net. At the same time, the political and economic forces moving swiftly to regulate and commercialize the Net and the new social media are at least equally strong and may win out if left to their own devices. Although the possibility for cyberdemocracy is still there, it will not happen without considerable vision and effort on the part of those who want it.

Direct Bureaucracy?

"Direct democracy" in terms of "policy making" is only one part of the governance picture, though it unfortunately is the only part that most electronic democrats focus on. But how will the outcomes of "direct democracy" be fairly administered? If the answer is not "by direct bureaucracy," then all our efforts at direct democracy will probably come to naught, and bureaucrats will effectively rule.

Bureaucracy was itself once a great, progressive invention. It was created to counter the evils of what was called in the United States "the spoils system," by which the friends and families of victorious politicians were given jobs in the administration of the government. To prevent this, "a government of laws and not of men" was created, with a professional, faceless, unbiased, and almost inhuman bureaucracy. As a consequence, this once-great invention can, in fact, become a mechanism of control by the ruler over the ruled – a form of power by which low-status people in boring, routine governmental jobs get to torture people who are trying to obtain a legitimate service. If those are some of the major functions and pathologies of modern bureaucracies, then what should design objectives be in relation to them? Popular proposals currently say: reduce/eliminate the mechanism of control and the form of power functions; see that lobbying is done conscientiously and fairly for all; turn bureaucrats into a kind of ombudsman or public defender; outsource everything possible to the private sector. This last is currently dominant and has proven to be wholly unsatisfactory. This is because it makes a process already difficult to control democratically and quickly, almost entirely free of democratic influence of any kind.

A better option might be to develop and use more democratically transparent computer software systems for more administrative functions – for all routine, fair, equitable, quick, objective carrying-out of legislative/executive/judicial decisions. It is truly inhuman to expect people to behave like machines, but that is what conventional Weberian administrative theory intends. Once upon a time there was no choice. Turning humans into objective and fair automatons was our only option. Now we have machines that can be fair, objective, and quick beyond any human, no matter how bureaucratic he or she may try to be. Reduce the number of human bureaucrats to the very few who must actively interpret laws and act for constituents, recognizing that this then makes them an important part of the policymaking process as well. Finally, and most importantly of all, as with direct democracy, so also with direct bureaucracy: part of the duties of each citizen is not only to participate fully in making community decisions, but also in carrying them out. Every citizen must serve as an administrator for some portion of his or her life.

Direct Adjudication?

How can disputes and conflicts be resolved in a direct democracy? An answer is by direct mediation. As in direct democracy so also in direct mediation, seek first of all to create a situation where people solve their own conflicts among themselves rather than by hiring a lawyer to duke it out in a court of law. The overriding motto is: "Real men settle their own disputes peacefully among themselves. Real men don't hire a lawyer and go to court. And real men certainly don't use or threaten violence!" Mediation has risen in popularity in most common law jurisdictions as an alternative to the formal adversary system in many cases, reducing the adversary system to an exceptional last resort, or for use in truly exceptional cases. Many jurisdictions now require in certain situations that mediation be tried before a case can go to court.

Mediation is quite different from traditional judicial decision-making. The main point of mediation is to help the accused and the accuser reach an agreement that suits them, in contrast with one forced upon them by some distant judge applying some abstract law. A mediator is neither a judge nor an arbitrator. Judges and arbitrators enforce their (or the law's) judgments on disputants. A mediator helps the parties themselves reach an agreement that is mutually acceptable to each other. A mediator thus must be a skilled, impartial, patient facilitator.

In a democratic society, all people should be trained to be mediators from a very early age so as to settle their disputes peacefully among themselves. If that is not possible, then mediation lets people say what they want to say, while mediators strive to protect both parties from abuse by the other. In contrast with current formal judicial systems, anyone can participate and give their version of what happened to cause a dispute and what the accused and accusers are like. The process is interested mainly in restoring peace between the disputants so that the conflict won't fester and lead to another conflict. It is focused on the feelings of the disputants (and their families, friends, and neighbors) more than on the peace of the community generally, but has the effect of provoking both a sense of justice and of community tranquility.

Another possible judicial solution might also be via expert systems and artificial intelligence. Since much of dispute resolution is very routine decision-making, why not use expert systems and AI as much as possible here as well? Reduce the jurisdiction of a formal judiciary to only the most severe or complex cases that will be mediated by humans, if possible, or adjudicated by a judge, if necessary. Fully automated kiosk legal systems now exist in some jurisdictions where most disputes can be settled, with little or no cost, by answering a simple, quick, and easy-to-understand set of questions. At the least, one can do all the formal paperwork oneself to present to a mediator, or to a formal court for adjudication. Kiosks might initially be located in the lobbies of courthouses, but they should spread to supermarkets, malls – wherever groups of people normally congregate. Of course apps should be available on social media devices. The creation of electronic "virtual courthouses" remotely accessible synchronously or asynchronously from anywhere in the world is technically feasible now.

Beyond the "Three Branches"

So far this has all been really old-fashioned thinking based entirely on experiences on Earth. The idea of "separating" power into three distinct but overlapping "branches of government" was another great idea of the eighteenth century that may have run its course. There is much more to governing now than just making and carrying out formal laws and adjudicating disputes. It seems clear that the military of most countries is out of effective democratic control largely because so many military and paramilitary duties are now outsourced to private firms. Indeed much governance in all sectors is privatized, while the power of multinational corporations has grown beyond anything imaginable in the eighteenth century. Clearly democratic control of these corporations is necessary, if "democracy" is to have any real meaning at all. Mass and social media shape people's minds and behavior more powerfully than most mere laws can. The media also should be brought under democratic supervision as well not to suppress but to make real the vaunted rights of "freedom of speech" and "the press."

Freedom or Order?

But do we need formal systems of governance at all? Is self-governance (self-organization) possible? Or, more accurately, is it possible again since humans were self-governing without formal structures enforced by laws and police for most of our history in hunting and gathering societies? Complexity theory suggests that self-organization is possible without central control and is effective and efficient if we design well. There appears to be much that humans can learn about governance from the behavior of swarms of certain animals.

However, the reactions of many people, and not only in the United States, to the events of September 11, 2001, suggest that many people do not want "self-organization" or even "democracy." In spite of repeating the old slogan of the American revolutionary war period, "Give me liberty or give me death," when actually given a choice a surprising number seem to have said instead: "Give me shelter." Far from being stalwarts in "the land of the free and the home of the brave" Americans, and many others, seem instead to be easily frightened, desiring security over liberty.

Indeed, political philosophers have long said that one of the major tasks of a well-designed governance system is to balance the opposing desires for both freedom and order. Too much freedom brings calls for order, and too much order brings calls for freedom, and so the pendulum moves back and forth between the two, more typically staying on the side of order longer than on the side of freedom. There are many other related dichotomies, too, that constantly need to be readjusted – the individual versus the community, centralization versus decentralization, local versus global, and many more. In need of special consideration is the question of "will" versus "structure." Should we try to make people want to do the "right things" via laws, education, religion, and exhortations, or can we make people do the right things regardless of what they believe by creating structures, perhaps even physical structures, that make it easy to do the right thing but very hard to do the wrong thing? At present, we typically tell people that they should do the right thing but make it very difficult or costly for them actually to do so because of the way society and the environment is structured.

Newton and Newtonian social design emerged from period when the West was rising to global cultural dominance. Thus all governance design and its alternatives so far have come from western cosmologies and politics. What would designs based on other cultural cosmologies be like? What if we had founding mothers instead of founding fathers?

Governance for Not-Earth: Back to Basics

All governance systems so far have been for Earth with Earth's biosphysical features. No place in space has those features. Many things that we take for granted on Earth, such as gravity, oxygen, atmosphere, trees, and rivers, do not exist in not-Earth. Instead, we will be required to create and maintain things in space if we believe we need or want them. Imagining, inventing, and maintaining these things we simply take for granted on Earth may well be the main governance challenges of space.

Ideally, of course (and probably eventually if not initially) spacekind will adapt to the environments of not-Earth and let go of the Earthly things that no longer make sense. One of those things might be the historical conventions of governance and the current trappings of the state. These are: (1) the tripartite separation of powers; (2) written constitutions and laws; (3) a small number of "leaders" and a large number of "followers"; (4) representatives; and (5) a criminal justice system. Governance designers will simply ask, generically, "What needs to be done; who/how shall it be done; and what happens if it is not done well/properly?" and design on the basis of that.

Examples of things that might "need to be done" that might be part of governance design include:

- Deciding policies for the community;
- Carrying out those policy-decisions;
- Preventing/minimizing/resolving interpersonal/organizational conflicts;
- (That is the tripartite division of government, of course.)
- Defining and then balancing the dichotomies listed above and others;
- Dealing with issues between unaugmented humans, cyborg-genetically-modified humans, transhumans, post-humans, cyborgs, intelligent robots, and pure artificial intelligences;
- Caring for the young, the old, and the frail;
- Creating and caring for the 'natural' and the 'built' environment;
- Socializing the young and newcomers into group norms;
- Requiring/permitting/forbidding religious beliefs and activities;
- Punishing/correcting/tolerating/enjoying/being deviants;
- Creating and disseminating (and/or withholding and censoring) ideas, data, and information;
- Creating or enabling recreational, entertainment, and/or sports facilities/activities;
- Producing and distributing food, goods, and services;
- Providing/maintaining private and/or communal habitats, communication and transportation systems, pure water and clean air, energy, and waste disposal;
- Preventing or mitigating disasters and emergencies;
- Relating one's settlement to other settlements on Mars, between Mars and Earth; and between Mars and other space settlements;
- Interpreting the meaning/intent of the governance design;
- And more.

The governance design may specify that

1. None/some/all of these functions must be collectively decided upon and/or carried out, or
2. None/some/all must be left to "the market," or
3. To individual initiative, or
4. To some mix between those three alternatives – or
5. There may be completely different ways of conceptualizing the matter.

The purpose and location of the settlement also affects the preferred design. Here are seven alternatives for a Mars settlement, each probably preferring governance systems different from the other six:

1. Established by the European Space Agency alone, funded by the EU alone. Purpose: minimally intrusive scientific exploration of Mars.
2. Established by NASA alone, funded by the U. S. government alone. Purpose: to terraform Mars as quickly as possible.

3. Established by former and current students of space universities and programs worldwide, and funded by a consortium of non-profit and space-oriented entrepreneurs and nongovernmental organizations worldwide. Purpose: to create an international university on Mars.
4. Established by a multinational mining corporation. Purpose: mine Mars for resources for further space exploration and exploitation.
5. Established by a global, "new age" spiritual group. Purpose: to create a community where they can practice their unpopular and unconventional spiritual beliefs freely.
6. Established by a consortium of wealthy entrepreneurs. Purpose: to create a popular resort for tourists (perhaps also sex tourists) from Earth.

When asked to design new forms of governance without merely adapting current forms, many people end up favoring one of two basic designs. Some try in effect to reinvent governance systems of functioning hunting and gathering societies. They feel that most people want to be able to participate directly and easily in matters affecting them. They want to have a fair hand in carrying out group tasks. They want to participate in settling conflicts among themselves without judges or bureaucrats or rulers. However, they understand that few people want to be fully occupied in political decision-making all the time. There are many instances when people are happy to have other people make decisions for them, as long as they can weigh in on the things that really matter most to them whenever they want to.

To achieve this, designs focus on having people live in small face-to-face communities where they can govern themselves and carry out their decisions directly most of the time while also having an elected representative (whom they may replace at any time, for any reason, and participate directly themselves if they wish) to make decisions on their behalf otherwise. They typically solve the problem of scaling up to large populations by having elected representatives from their community who then meet with other community representatives, and so on up the line to whatever the largest assembly of representatives may turn out to be. Most realize this is not an entirely satisfactory solution – the farther away the representatives are, the less likely people in the local communities are to be satisfied with the decisions – but they seldom can come up with a solution.

On the other hand, there are many people who would choose to leave all the governing (and much of any other work needed to be done, including law enforcement) up to "computers," or artificially intelligent robots, and the like. They assume that these entities will be omnipresent, omniscient, and unbiased, rendering solutions and performing actions that are best for everyone. Humans thus are free to play and pray as they wish without either working or governing themselves, although somehow they imagine humans can, if necessary, intervene if the machines go wrong. Some worry a lot about that.

These two basic patterns have many specific variations. And there are other solutions as well. Dictatorship, normally an intelligent, honest, and benevolent dictatorship (Singapore is often referenced or a captain of a ship) is popular especially for people who recognize how precarious not-Earth will be for Earthlings, at least at first. Order must take precedence over freedom, they feel. Others, on the contrary, either have faith in humans to do the right thing if they feel directly and immediately the consequences of their acts, or believe in the superior wisdom of mobs and swarms. They propose complete anarchy, with "whatever has to be done" being done by whoever wants to do it since the settlement will fail if those things are not done. Or they favor assigning all duties by lot, and reassigning them frequently, and not by election – certainly not by majority rule, which only makes opponents out of losers – or by consensus, which takes too long to do and can be endlessly thwarted by sociopaths.

Our Best Hope for Better Governance in Space?

More than anything else, space is, or should be considered to be – or should be understood that it will become – a place to do things that cannot or have not yet been done on Earth. So far, it seems that insufficient thought has been given to why people might choose to go into space to live; how they might like to behave in not-Earth (perhaps in contrast to the way they are required to behave on Earth); what the experience of living in not-Earth itself might do to them, individually and socially; and how they might govern themselves during the transformation as well as after. Aspirational and new governance issues are very rarely discussed in the formal, "scientific" literature of space exploration and settlement (of course it is found in some, but not even most, space fiction).

Unfortunately space law based on Earth practices has overwhelmed space governance. Although there are academic centers and journals devoted diligently to seeing that current Earth law expands into space, few efforts have been devoted to imagining and designing completely new systems of governance for twenty-first century space settlements and beyond. But why would anyone wish to settle in space if not to live better lives than they can on Earth? Imagining and creating new governance systems, based on new ideas, aspirations, and cosmologies, should become the center and purpose of all space studies and activities.

References

T. Becker, C.D. Slaton, *The Future of Teledemocracy* (Praeger, Westport, 2000)

M. Daigle, Dynamic democracy (2011), http://upgradedemocracy.org/

J. Dator, Designing governance systems for Mars, in *Space Travel and Tourism: New Frontiers*, ed. by R. Harish (Icfai University Press, Hyderabad, 2007)

C. Gray, *Citizen Cyborg: Politics in a Posthuman Age* (Routledge, New York, 2001)

J. Hughes, *Citizen Cyborg. Why Democratic Societies Must Respond to the Redesigned Human of the Future* (Basic Books, New York, 2004)

M. Mannermaa et al., *Democracy and Futures* (Parliament of Finland, Helsinki, 2006)

F. Riggs, Presidentialism: a problematic regime type, in *Parliamentary vs. Presidential Government*, ed. by A. Lijphart (Oxford University Press, Oxford, 1992)

G. Robinson, Space law for humankind, transhumans and post humans – is there a need for a unique theory of natural law principles? Annals of Air and Space Law **332**, 287 (2008)

G. Robinson, H. White, *Envoys of Mankind: Declaration of Principles for the Governance of Space Societies* (Smithsonian Institution Press, Washington, DC, 1986)

Chapter 8
Do Rocks Have Rights?

Jacques Arnold has an interesting job within CNES, the French space agency. He has been a Space Ethicist with them since 1998. His job is to talk with people working at CNES about the ethical implications of their work. He does not believe that ethics is a matter of a set of rules of behavior that must be followed and enforced, but rather a way of thinking and acting reflectively and responsibly in everything one does. He helps people think through the ethical consequences of their work in many dimensions, perhaps leading them sometimes to do things differently from the way they otherwise might have done them.

Although some other space agencies have occasionally thought about the ethical implications of their activities, and while UNESCO for a while had a special focus on space ethics, no other space agency has been as consistently concerned about ethics as has CNES. Arnold is a perfect person for the job. He is a tall and imposing presence on the one hand and a gentle, kind, soft-spoken, and reflective person on the other. He has a way of drawing people out so that they themselves become practical ethicists as a consequence.

Why should ethics be of particular concern to space agencies and industries? Of course, ethical behavior should be of major concern to all of us in all occupations, but because of the special nature of space activities – their expense, their danger, their public visibility, and their unique challenges and opportunities – space and ethical reflection must go together.

Ethics Versus Morals

There is a distinction between ethics and morals. Ethics concerns "doing the right thing." Morals concern "doing the good thing." Ethics are admittedly human-made and can thus be changed by humans. Morals typically have a religious base, and thus from the point of view of the religious believer come from God or some other higher authority and cannot be changed by humans. Many religious people assume

J.A. Dator, *Social Foundations of Human Space Exploration*, SpringerBriefs in Space Development, DOI 10.1007/978-1-4614-3094-0_8, © James A. Dator 2012

that people who don't believe in God will not behave properly. All people need divine, moral guidance. And yet there are many cultures – Japan is often given as an example – where human relations are exemplary even in the eyes of religious believers, even though their behavior is based entirely on cultural ethical standards and not on divine moral requirements. Japan was once called "the missionary's graveyard" because there were so few sinners needing conversion. Indeed some argued that the behavior of many Japanese was superior to that of some of the missionaries.

Ethics can be absolute (the rules apply to everyone everywhere) or relative (depending on the situation and/or people involved). Morals tend to be absolute, demanding adherence under all situations and by all peoples everywhere.

This chapter is about space ethics, and not about space morals. For the latter, we would need to go into much more detail about what various religions have to say about human space activities (including whether humans should dabble in space at all). In addressing morals one has to consider what behavior is expected of humans in space in relation to each other (and what morals apply to those of the same or different or no religious persuasion), let alone any extraterrestrial life humans might encounter in space. Finally in the realm of morality one would need to ponder over whether it is moral for humans to exploit and "develop" extraterrestrial environments for human benefits, or whether those environments have inherent values that should limit or prevent human intervention. We will not consider those issues here from a religious or moral perspective, but we will consider such issues from an ethical perspective. Thus we start by clearly admitting that ethics, being human made, are limited, partial, and perhaps tautological since it is humans ourselves who say there are or are not certain limits to human activities in not-Earth.

Some Ethical Challenges

Ethics has many levels or dimensions. At the most basic level, it is unethical to cheat, lie, not do your job as you are supposed to do, mislead people – for example, saying you have tested something you were supposed to test, but not testing it, or testing it superficially. This kind of unethical behavior has had serious consequences in the aviation industry when mechanics and their supervisors have said they did things they did not do, causing a plane to crash with loss of life. The consequences could be even more catastrophic in space industries, and indeed there have been ethical questions raised about many of the space disasters that have occurred.

It is clear that behaviors such as the ones just listed are unethical. But sometimes, matters are not so clear cut. An instance of an ethical dilemma is telling the truth but not the whole truth in contrast to telling the whole truth when a partial truth would be OK and the whole truth would be unnecessarily damaging. On the one hand only "answering the question" instead of telling the person what he or she really wants to know can appear to be ethical, but not if the partial truth leads to misunderstandings and tragedies. On the other hand, most people agree that there are times when it is best not to tell someone "everything" you know about something, since it will only hurt their feelings and not lead to any superior consequences. Balancing these two can be difficult. It is easy to err here.

Special challenges in the space industry are the ethical obligations that experts have towards non-experts, i.e., the public. Very often, a "truthful" explanation of a scientific or technological process is so complex and requires so much prior knowledge that the expert simplifies so much that she "lies" or else uses metaphors that the non-expert can grasp but often in a way that miss the essential point of the scientific/ technological issue. There are many examples in the attempts to explain quantum physics to laymen, resulting in metaphors that probably give non-specialists erroneous understandings of the principle. This understandable desire of experts to do good can also lead to having an expert "help" others who don't want to be helped, or causing people to think or act as you want them to act because you think it is better for them. This is often the essence of ethical dilemmas facing (but seldom admitted by) policymakers, religious missionaries, advertisers, teachers, and perhaps even your mother. Is it ethical for Christian missionaries to destroy highly functional and long-established native cultures and practices in order to "bring them to Jesus?" Obviously the missionaries think that is a no-brainer, while religious skeptics think it is a no-brainer, too, but from an opposite point of view about what is correct behavior.

Far more common are doing things that may hurt others but will profit yourself. When that is phrased abstractly it would clearly seem to be unethical. But our recent economic system was largely based on this dubious ethical proposition and widely defended by the slogan derived from the movie "Wall Street" that "greed is good" – a necessary element in order to motivate people to innovate, invent, and improve life for everyone by allowing some to become extremely rich and powerful while others are (temporarily?) impoverished. It has often been said that there is a thin, almost indiscernible line between a shrewd businessman and a crook.

A particularly vexing corollary of this is our willingness to "throw away" our waste – out the window of our speeding car, into a landfill, or in the case of the space industry, into the "vastness" of space in the belief that it will eventually be burned up as gravity pulls it back to Earth, and that it will not endanger other space objects until then. The consequences of this very unethical though extremely common (and often economically justified) behavior is now so serious that current space activities are endangered and, unless our junk is cleaned up and further pollution prevented, may render some future space activities impossible. The lasting and damaging feature of space junk has been well known in the space community for many decades, but nothing effective has been done to eliminate or prevent it, in spite of many promises that are routinely overturned by "economic" or "military" justifications – or by simple individual acts of unethical carelessness.

Beyond "the Golden Rule"

This issue also segues into a much more perplexing issue: our ethical obligations to future generations, to the unborn whom we will never meet, and so cannot "get back" at us, but whose lives we influence by our acts today. We will discuss an aspect of this issue in more detail below, but one ethical argument made in favor of present generations accepting ethical obligations to future generations states that "Earth and space are not ours to exploit; they are treasures we must safeguard for our descendants." Needless to say, people who wish to exploit those "treasures" for immediate profit, as they do to the treasures of Earth, strongly disagree, and current laws and practices are on their side and not on that of future generations.

All traditional ethical systems are based on "reciprocity." That is to say, all ethical traditions have some version of what is often called "The Golden Rule," which is: "*Do unto others as you would have them do unto you.*" Or stated negatively: "*Do not do to others what you do not want them to do to you.*" That is "reciprocity." I can do to you and you can do back to me. So I should not do bad things to you *since* you can do bad things back to me. For most of human history that was a great rule. But it assumes that all humans like and dislike the same things. For most of human history, humans lived in small communities where that was true for them. But it is a big problem for us now who live in a globalizing world composed of many different cultures so that what one person may like another may hate, and vice versa. What is "clean" (or at least neutral) in one community may be dirty in another.

In some Pacific island communities it is perfectly OK to "borrow" a parked car for weeks at a time. Such islanders are shocked and outraged when they come to America and are arrested for "stealing." What a waste to leave a car parked when I need it to go somewhere! But in some of these same Pacific island communities it is absolutely not OK for females to show bare thighs. Thus some islanders are offended by Western women in t-shirts and short shorts, while Westerners may be offended by island women going topless in long skirts. In some cultures it may be wrong to talk about or even think about the human body at all. So, in our global world, the Golden Rule often doesn't work. A better ethical rule now may be: Do unto others as *they would have you do unto them.* That is, take the time to find out what is pleasing and repulsive to others, and try to please and not repulse them. That may be the more ethical thing for you to do – for them – but very hard for you to do since you may have to violate your own ethical standards!

However, the Golden Rule also assumes that if I do something to you that you don't like that you *can* do something back to me to make me aware of that. So the rule is good for a local community. But we don't worry if what we do impacts people on the other side of the world, since they cannot get back at us. Or at least we don't worry about them until they can and do get back at us. Most Americans didn't know or didn't care that some of their actions injured or offended some people far away, until these people got our attention in a very spectacular way on September 11, 2011. Since then, Americans have been perhaps overly concerned about offending them again – or rather, about making sure they can't get back at us again.

There are other things that earlier systems of ethics did not have to address. A new ethical challenge for all people in the so-called developed and developing parts of the world now is the impact they have on future generations. For almost all of human history, doing what was right for present generations was probably also doing what was right for future generations. But over the last several hundred years, and especially the last 100 years, present generations have been able to do things that have very powerful and lasting effects for many generations beyond – both giving them new burdens they must bear for a very long time and using up resources they might want to use themselves. So we can *do* to them, but they can't *do back* to us. Present generations can and often do seriously impact the lives of future generations, but future generations cannot get back to thank or punish us. A new ethical obligation of present generations towards future generations that is not based on reciprocity must be recognized, and new institutions created, to see that future generations can somehow be "present" when we make decisions impacting them.

New Rights for New Beings

The history of rights (and ethical obligations) has been an evolving one. (We are thus now switching from "ethics" to the language of "rights" here. A "right" is a legal concept, but rights are often based on prior ethical expectations). For most of civilization (i. e., the last 5,000–8,000 years), it was typical that only the lives of a small elite were important enough to require ethical treatment. It was, for example, perfectly OK for white men to ignore women. They were deemed to be inferior to and dependent on men – not fully human. Only recently have women become human enough to deserve equal ethical consideration, though the struggle still continues.

So also with children's rights. What is now considered to be child abuse in many postmodern societies was normal, expected childrearing practice a generation or two ago. "Spare the rod and spoil the child," it was said. "A child is to be seen and not heard" was another statement. Children expected to be beaten by their teachers or principals, and certainly by their fathers, if they "acted up" or "talked back" to adults. Now that is considered to be child abuse. Ethical norms have changed.

Similarly, married women were in a legal sense "owned" by their husbands until the 1960s in America. Wives did not have the right to go into debt, i.e., to own a credit card of their own in their name. Their husbands had to give the card issuer their permission. And although it was not exactly OK for a man to beat his wife "if she needed it," neither was it OK for her to become "uppity," to neglect her wifely and motherly duties, and to put herself "above" her husband and children. What was accepted practice a few decades ago is now considered to be spousal abuse. Ethical norms have changed.

So also with emerging animal rights. Anyone who follows what Americans call "football" knows what happened to a highly paid athlete who ignored that it is now considered unethical to torture dogs for human enjoyment. But, again, such behavior was not uncommon a few years ago. Moreover, almost no one took a sick dog or cat to the veterinarian except in rare situations, and either they or the vet would quickly kill the animal rather than pay huge sums of money for operations or prostheses. Now, pets are intimate members of many families who we love and care for very deeply, perhaps enjoying better health care than the children, and sometimes inheriting wealth when the owner dies. Ethical norms have changed.

Robot Rights?

And now, here come artificially intelligent autonomous beings, perhaps. What about the rights of robots? Robots were originally created to be dumb, fully programmed entities, built and expected to do dirty, boring, and dangerous jobs so that humans – even slaves or women – need not do them. Like women and slaves before them, robots currently have no rights. Humans have no ethical obligations towards them. Unplug them when we want to. Throw them away when they are finished or obsolete. But, as we discussed in Chap. 6, it appears that truly intelligent and sensitive robots are emerging. Moreover, even now robots are being made to resemble humans physiologically, and to be very attentive and responsive to the emotions and feelings of the humans around them. They are being made to appear they care about us. This most certainly will create even stronger emotional bonds of humans towards robots. And as robots become truly intelligent and sensitive, they may begin to demand to be treated ethically – according to their ideas of what is ethical, not necessarily by our standards.

Do Rocks Have Rights?

Finally, do rocks and trees have rights? That is, do landscapes, and even abiotic natural formations have the right to be left alone? Do ethical obligations only apply to living things or sentient creatures? More and more people are beginning to argue that landscapes, environments, and other objects in nature should command ethical considerations from humans. Indeed, more and more people are beginning to understand that entire biospheres have rights – or that there should be limitations to what humans can do to a biosphere and its parts. Our failure so far to grant rights to trees and rocks may be leading to the extinction of humans, whose existence is dependent on the existence of trees and rocks. There may be reciprocity at a planetary scale going on now on Earth. We have done it to nature so nature is doing it back to us.

Obviously everything discussed so far has relevance as to whether we should terraform Mars, or even whether humans should land on the surface of any object in space until we know a whole lot more about what "life" is and the composition of space objects. More people are saying: we really should not "boldly go," but go cautiously, sensitively, and ethically into space. "First People" are said to ask the permission of a deer before they kill it for food and materials, and many people "say grace" before eating anything, thanking God and nature for providing the sustenance humans need. So also should humans humbly ask permission of Mars and the asteroids before we terraform or exploit them, if we should exploit them at all. Many other people argue this is ridiculous; that Earth and the universe were made for humans to exploit and we should do so without the slightest qualms of conscience. People who say Mars or other objects in space have "rights" or require us to behave "ethically" towards them are probably people or institutions who have no means to get into space to exploit it. Some people suggest that this is perhaps a diversion. They suggest that such cautions about mining the Moon and Mars are only said so they can catch up and mine it themselves! It is only "backward" countries and agencies that worry about ethics. What they are really doing is trying to thwart the great American entrepreneurial spirit of free enterprise – until they can catch up.

Nonetheless, will humans be deterred in our desire to exploit space for economic, military, nationalistic, or other purposes by ethical concerns about the rights of rocks, trees, or robots? The entire history, present, and probable future of junk on Earth as well as in space strongly suggests that humans only feel regret over the consequences of our stupid acts. We seem unable to exercise ethical foresight before we act regrettably once again.

References

J. Arnould, Space, ethics and society. A CNES study. Acta Astronaut. **48**(5–12), 917–920 (2001)

C. Cockell, The rights of microbes. Interdiscip. Sci. Rev. **2**, 141–150 (2004)

M. Coeckelbergh, Robot rights? Towards a social-relational justification of moral consideration. Ethics Inf. Technol. **12**, 209–221 (2010)

E.M. Daly, R. Frodeman, Separated at birth, signs of rapprochement: environmental ethics and space exploration. Ethics Environ. **13**(1), 135–151 (2008)

P. Lin et al., *Autonomous Military Robotics: Risk, Ethics, and Design* (US Department of Navy, Office of Naval Research, Washington, DC, 2008)

M. Lupisella, J. Logsdon, The rights of Martians. Space Policy **13**, 89–94 (1997)

A. Pompidou, *The Ethics of Space Policy* (UNESCO, Paris, 2000)

W. Wallach, C. Allen, *Moral Machines: Teaching Robots Right from Wrong* (Oxford University Press, Oxford, 2009)

Chapter 9
If They Are, Where Are They?

Serious people have speculated for eons about the possibility of life and intelligence in not-Earth. Democritus (Greece, 460 BCE-370 BCE) said with amazing foresight: "In some worlds there is no Sun and Moon, in others they are larger than in our world, and in others more numerous. In some parts there are more worlds, in others fewer (…); in some parts they are arising, in others falling. There are some worlds devoid of living creatures or plants or any moisture."

Lucretius, an influential Greek philosopher of the first century, BCE, also commented quite logically: "If the fund of seeds is so vast that the sum of the lives of all living creatures would not suffice to count it, if the same force of nature is still operative and possesses the power to assemble all the seeds of things in the same order in which they have been assembled in our world, you are bound to admit that in other parts of the universe there are other worlds inhabited by many different peoples and species of wild beasts." (Lucretius, *On the Nature of Things*) The Roman poet Cicero speculated on the possibility of living beings on the Moon, and Plutarch wrote an account of a visit to the Moon. As we saw in Chap. 2, Lucian of Samosata wrote of similar adventures in his *True History*.

Nonetheless, at the same time, equally – perhaps greater – minds of the same period, for example Plato and later Aristotle disputed the possibility of life in not-Earth, arguing that the existence of many worlds would destroy the harmony of the created world that placed Earth at the center of the universe. On the basis of meticulous observations and logical arguments, Claudius Ptolemy refined the Aristotelian model and for the next 1,000 years, little was heard in the West of ideas of life in not-Earth. The Aristotelian/Ptolemaic view prevailed – and worked amazingly well.

However, on the other side of the world, Teng Mu, a Chinese scholar of the Sung dynasty (thirteenth century), came to the same conclusion as Lucretius: "Empty space is like a kingdom, and earth and sky are no more than a single individual person in that kingdom. Upon one tree are many fruits, and in one kingdom there are many people. How unreasonable it would be to suppose that, besides the earth and the sky which we can see, there are no other skies and no other earths."

J.A. Dator, *Social Foundations of Human Space Exploration*, SpringerBriefs in Space Development, DOI 10.1007/978-1-4614-3094-0_9, © James A. Dator 2012

Others around the world came to similar conclusions. In the middle of the sixteenth century, Copernicus, in *De Revolutionibus Orbium Caelestium,* argued convincingly for a heliocentric Solar System that contradicted the dominant religious dogma of the time, which endorsed Ptolemy. Copernicus inspired Giordano Bruno (1548–1600) to entertain many heretical beliefs, including the possibility of many worlds with many forms of life in the universe (for which Bruno was eventually executed by the Church), while with the improved telescope Galileo (1564–1642), also persecuted by the Church, saw mountains on the Moon and eventually the four largest moons of Jupiter.

Johannes Kepler (1571–1630) showed that planets move in ellipses rather than the perfect Ptolemaic circles and also wrote a fictional story of a trip to the Moon. Shortly thereafter, Bishop Francis Godwin wrote a similar story in English (1638). The British philosopher/poet Henry Moore declared, "But if that infinite Suns we shall admit, / Then infinite worlds follow in reason right, / For every Sun with Planets must be fit, / And have some mark for his farre-shining shafts to hit …" (1647). Bernard le Bovier de Fontenelle's *Conversations on the Plurality of Worlds* (1686) appeared as a conversation between two people walking late at night, contemplating the stars. It was a strong defense of the Copernican worldview focusing on the possibility of extraterrestrial intelligence.

New technologies encouraged people to challenge established cosmologies while reaffirming neglected ones: "All these illustrious worlds, and many more, / Which by the tube astronomers explore: / And millions which the glass can ne'er descry, / Lost in the wilds of vast immensity; / Are suns, are centres, whose superior sway / Planets of various magnitudes obey…. / We may pronounce each orb sustains a race / Of living things adapted to the place. / …. / How many roll in ether, which the eye / Could n'er, till aided by the glass, descry; / And which no commerce with the Earth maintain! / Are all these glorious empires made in vain?" (Richard Blackmore 1712).

Communications Between Earth and Not-Earth

There soon was a swelling flood of stories about contact between humans and not-humans from not-Earth. Eventually some practical as well as visionary folks dreamed up ways to let intelligences on not-Earth know that some Earthlings were ready to talk with them. Apparently one of the first was the great Russian mathematician, Karl Gauss, who proposed in 1820 to create giant right triangles out of wheat and forest growth in Siberia that, when viewed from space by any inquisitive aliens, would clearly depict the Pythagorean theorem, demonstrating that intelligent life existed on Earth. Joseph von Littrow had a similar plan in 1840, proposing cutting huge ditches in the Sahara desert, filling them with oil, and setting them on fire. The French physicist, Charles Cros, proposed in 1869 that giant mirrors be placed across Europe with the Sun's rays reflected towards Mars so as to attract the attention of Martians living there.

These experiment were never undertaken, but were among the many ways in which people have tried to communicate with aliens by using whatever technologies were available at the time. We saw earlier that some ancient stone structures and markings in Earth have been interpreted as messages to not-humans in not-Earth. So when radio came along, it was inevitable that clever folks would devise ways to use radio to send and receive messages from ETs. Nikola Tesla tried to send radio signals to ETs and at one point thought he heard a signal from ET that turned out to be a natural sound called "whistlers." Marconi himself listened for ET communications. Since radio was then the most advanced communication technology, why not assume that ETs would use it to try to find or contact humans via radio?

SETI: The Search for Extraterrestrial Intelligence

The current era of SETI began with a paper published in 1959 in the prestigious British magazine *Nature*, entitled "Searching for Interstellar Communications," by Philip Morrison and Guiseppe Cocconi. Frank Drake also used the radio telescope of the National Radio Astronomy Observatory to look for signals near two sun-like stars. Many other low budget or unfunded observations in the United States followed suit with no success. There was considerable interest in SETI in the Soviet Union at this time, and several books, some involving U. S.-USSR cooperation, also appeared, such as Walter Sullivan's *We Are Not Alone* (1964) and L. S. Shklovskii's and Carl Sagan's *Intelligent Life in the Universe* (1966). One of the more interesting and prescient pieces was *The Cyclops Report* in 1971, edited by NASA-Ames director John Billingham and Bernie Oliver, director of Hewlett-Packard Research Labs. The Academies of Science in the United States and the USSR jointly sponsored a meeting in Byurakan, Armenia, also in 1971. The International Academy of Astronautics created a SETI committee. The serious, scientific, legitimate search for intelligent extraterrestrial life appeared to be well underway.

NASA provided funding for exploratory SETI workshops over the 1970s until Senator William Proxmire bestowed one of his "Golden Fleece" awards on NASA in 1978, as he did on so many creative and controversial activities: he declared to be a waste of taxpayer's money. Interest in SETI continued to grow nonetheless, and Congress once again briefly provided funding for a promising microwave observing SETI project. However, in 1990, Congressman Ronald Machtley said, "We cannot spend money on curiosity today when we have a deficit." Congressman Silvio Conte said there was no need to spend millions of dollars looking for extraterrestrial intelligence "when you could find the evidence in any tabloid at your local supermarket for 75 cents."

NASA's SETI program was finally killed in 1993, with Senator Richard Bryan saying, "This hopefully will be the end of Martian hunting season at the taxpayer's expense." Fortunately, funding for the search for extraterrestrial intelligence was provided by two private organizations, the SETI Institute and the Planetary Society. Today many individuals have allowed their idling PCs to be linked together by a screensaver called "SETI@home" to expand the search in order to process incoming signals relayed from radio telescopes that are monitoring the skies for detectable signals from space.

One of the important recent developments that should impact SETI is the systematic search for extrasolar planets that might contain life so that the search can be narrowed to a few prospects instead of scanning widely over the vastness of space. Since the first extrasolar planet was discovered in 1992, over 700 more have been discovered, with thousands more probably observable, so far all too far away to be anything other than possibilities for life. As the search continues, and our ability to focus on planets improves, clear candidates for closer inspection will likely emerge. The Hershel telescope deployed recently has been particularly adept at locating these so-called "exoplanets".

Protocols for Communicating with ETs

Most of SETI so far has been devoted to trying to identify evidence of intelligence elsewhere in the cosmos. But there also has been some attention devoted to what to do if a signal is believed to have been received. Jill Tarter (one of the most important figures in the SETI community) and others have developed a "Declaration of Principles Concerning Activities Following the Detection of Extraterrestrial Intelligence" in cooperation with the International Academy of Astronautics and the International Institute of Space Law.

The declaration has been adopted by several international astronautical agencies and is also on the web page of the SETI Institute. It is only a document for discussion and does not have any legal status whatsoever. The nine points of the Declaration of Principles propose that any believed signal or other evidence of ET should be carefully verified and re-verified by qualified scientists before the discovery is released to several national and international bodies listed in the declaration; the discoverer should then be the one to release the evidence and other necessary data to the broader scientific community and general public; the evidence should be carefully preserved and the channel through which it was received protected so as to receive more signals; no response to a signal should be given until the signal has been studied and a response officially authorized by appropriate international bodies; and the principles should be continually reviewed and upgraded as necessary.

Much more controversial are activities or proposals to send messages from Earth to ET. Nonetheless, the first spacecraft expected to leave the Solar System displayed messages composed for any ET who might encounter them. Plaques designed by Frank Drake and Carl and Linda Sagan were attached to the Pioneer 10 and 11 spacecraft (1972–1973). The most dominant features are sketches of a nude human male and female with the spacecraft in the background. Below them was a series of circles and quasar locations as seen from Earth, intending to show where the spacecraft originated, and a symbol for hydrogen, the most comment element in the cosmos.

For *Voyager 1* and *2,* launched in 1977 and also headed toward deep space, Sagan and others developed a far more complicated suite of natural and artificial sounds and images including greetings in 55 languages, and a mechanism and illustrated instructions for playing it all back.

In 1974, the huge Arecibo Radio Telescope in Puerto Rico sent a pictorial message to the star cluster M13. The picture contained 1,679 bits arranged into 73 lines of 23 characters per line. It consisted of crude sketches of the Arecibo telescope, our Solar System, DNA, a stick figure of a human, and signs of the biochemicals of Earthly life. In 1999, 2001, 2003, and 2008 radio messages with varying structures suggested by the Russian astronomer Aleksandr Zaitsev were sent from Ukraine to stars 20 and 69 light-years from Earth. The signal in 2008 included digital text messages, photographs, and drawings from ordinary citizens. Few people have been able to decode these messages without cues from their creators, and it is highly unlikely that even the most intelligent of ETs can decipher them without extensive prior knowledge of Earth and human culture. We'll see.

Allen Tough and more recently Dimitra Atri, Julia DeMarines, and Jacob Haqq-Misra (among others), have proposed that we think more carefully before actively messaging to ETs. We need to find non-anthropomorphic signals and both simpler and more sophisticated technologies. Should we assume that ETs can see, hear, and otherwise sense in the way humans can? We need to devise messages that do not presuppose human abilities only. Allen Tough says we should (1) pursue a variety of means for searching the Solar System and Earth for physical evidence of an alien object or its effects; (2) issue invitations asking ETI to have a dialogue with humankind; and (3) become sufficiently prepared for contact, thus encouraging ETI to respond. For evidence from many light-years away, Tough says the most promising strategies are (4) a search for astroengineering projects by ETs and their byproducts, and (5) radio and optical SETI. Tough created a web page specifically asking ET to call Earth – another example of assuming ET is currently using the same new technologies that humans currently are.

Discussing what they call "METI" – "messaging to extraterrestrial intelligence" – Atri, et al., say that any "protocol for METI should provide guidelines for the message length, signal encoding, transmission method, and information content in order to maximize the likelihood that the message is understood. Successful messages will minimize any anthropocentric or culturally dependent content and avoid dependence on basic human senses. Additionally, a message that is expected to be decoded and comprehended by extraterrestrials should be decipherable across human cultural boundaries on Earth." A very difficult standard to meet! As with receiving messages, so also with sending them, Atri, et al., say there should be standard protocols, accepted after worldwide discussion, in order to send ET a clear message and not a myriad of solitary, confusing signals as we currently are doing. Messages to ET should be short, simple (in terms of both mode of communication and message), and repeated over and over at regular intervals to assure ETs that it is not a fluke or of natural origin but a message intended from us to them.

The Drake Equation and the Fermi Paradox

By why even bother? Why should we assume there are any ETs able to receive our messages or to send some to us? One answer lies in what is known as the Drake equation, devised by Frank Drake for discussion at a meeting of the U. S. National Academy of Sciences in 1962. The Drake original equation states that:

$$N = R^*. fp . ne . fl . fi . fc . L$$

where:

N = the number of civilizations in our galaxy with which communication might be possible;
R* = the average rate of star formation per year in our galaxy;
fp = the fraction of those stars that have planets;
ne = the average number of planets that can potentially support life per star that has planets;
fl = the fraction of the above that actually go on to develop life at some point;
fi = the fraction of the above that actually go on to develop intelligent life;
fc = the fraction of civilizations that develop a technology that releases detectable signs of their existence into space;
L = the length of time for which such civilizations release detectable signals into space.

There have been many suggestions for adding elements to or otherwise modifying the equation. There also have been many numbers suggested for each of the elements of the equation. Some suggest a low probability of currently contactable ETs. But enough influential astronomers have suggested numbers that lead them to conclude, as Carl Sagan all-too famously said, that there are "billions and billions" of stars in the universe and many of these could support intelligent life. This belief is what drives all SETI activities now. On the other hand, as Enrico Fermi also famously asked, in effect, "If they are, where are they?" Why have no ETs contacted us and why have we found none or any evidence of their existence in our searches? This has become known as the Fermi paradox – the gap between the huge numbers of ETs believed to exist and the fact that we have not clearly found even one. Arthur C. Clarke once said on the subject: "Sometimes I believe we are alone in the universe, and sometimes I believe there is intelligent life elsewhere. In either circumstance, the implications are staggering."

SETI, Astrobiology, and UFOs

It is important to understand that the search for life and the search for intelligence in not-Earth are not the same thing. They each involve different theories, methods, and people. Some who believe we should look for extraterrestrial life do not favor looking for extraterrestrial intelligence. The search for extraterrestrial life is now often called "astrobiology" or "exobiology," terms that clearly privilege Earthlike life over other possible forms. Though it took a while to catch on in the scientific and political establishment, astrobiology now is generally considered to be a scientifically reputable enterprise. On the other hand, as we have shown, the search for extraterrestrial intelligence is way too controversial and so is undertaken, if at all, only by private, nonprofit organizations and not by governments. SETI becomes even more controversial when it segues into "ufology" – the belief that UFOs (unidentified flying objects) flown by ETs have visited us, recently and currently, and that some humans have been transported to UFOs, been examined by ETs, and returned to Earth. Ufologists often declare that there is a governmental cover-up of the existence of UFOs and so also sometimes blends into other conspiracy theories. SETI rejects UFOs. In this chapter, we are considering only SETI without coming to any conclusions about UFOs one way or the other.

So, does life exist anywhere in the universe except Earth? Does Earthlike life exist elsewhere? If not, is there non-Earthlike life out there? Could we recognize it, if so? Should we bother looking for it, or only concentrate on Earthlike life, as we are now officially doing?

Similarly, does intelligence exist anywhere in the universe except Earth? Does Earthlike intelligence exist elsewhere? What about non-Earthlike intelligence and our ability to recognize it if it does exist?

What Is Life?

One SETI project decided that living things have (1) structure and boundaries; (2) are in thermodynamic disequilibrium with their environment; (3) convert material into energy for themselves; (4) move; (5) adapt; (6) replicate; and (7) transfer information among themselves and perhaps with other life forms. Note that this definition tries to be universal and not unique to Earth, where definitions of life often say life is "carbon based" and "needs water and sun."

Nonetheless, some of these terms seem problematic. Is "replication" something an individual or a group does, or both? Is bisexual reproduction "replication"? Amoebae and other single-celled organisms reproduce asexually and thus "replicate." But do organisms that reproduce bi-sexually replicate? The offspring of bi-sexual reproduction come from and are related to their parents but certainly are not the same as either parent. Indeed it is the fact that offspring are different from parents that makes evolution via adaptation relatively speedy. Bi-sexual reproduction evolved comparatively recently on Earth. Accretion and asexual reproduction – both true replication – were the norm and are still widespread. Should we expect bi-sexual reproduction of ETs? Why not fission or tri-sexual reproduction?

Moreover, with bisexual reproduction came individual death. What is the point of living if we will die? Why not live forever? Crystals do. So might any ETs who might contact us. One of the biggest controversies about SETI and purposely sending a signal to ETs is the very widespread assumption that, being superior, ET would be at least as aggressive and willing to fight and kill as humans currently are, and with vastly superior technology to that of humans by which to fight and kill. But that assumes advanced life forms will live and die and hence use killing as a political weapon.

There is evidence well worth contemplating now that on Earth, "death is a curable disease" and that the human lifespan can be extended indefinitely. In part this is because in the future, perhaps no wounds or diseases will be fatal (all can be cured, or life restored). In part this may be because human intelligence that currently resides in our biological containers might soon be transferred to silicon or other containers that are more sturdy than frail human bodies. Humans, as transhumans or post-humans, might continue to live forever, intelligently, and with congruent memories of "their" individual or collective past. If we can imagine this with some level of realism on Earth, then surely we can assume that any ETs contacting us would be immortal and would have no desire to "kill" or conquer humans whatsoever.

A related question is whether time is real or is a byproduct of the birth, life, death trajectory of biology? If ETs (and we) can live forever, does our being "two million light-years away" matter to it or to us? We may all live in an eternal here and now.

What Is Intelligence?

Is intelligence the same as life? Is intelligence dependent on life for its existence? One definition says that intelligence is the ability to learn about, learn from, understand, and interact with one's environment; to understand and profit from experience; to adapt to new situations; to recognize that which is useful and that which is not. Another definition says intelligence includes the capacities to reason, to plan, to solve problems, to think abstractly, to comprehend ideas, to use language, and to learn. Those definitions suggest that intelligence is the same as life and thus don't seem very useful. A definition from animal intelligence says intelligence involves causal reasoning, flexibility (the ability to adapt to new situations), imagination, and prospection (the ability to anticipate, plan, and perhaps defer immediate gratification before acting). By this definition, many – maybe most – life forms seem to be intelligent.

What about the role of emotions? Humans are not always and in every situation rational. Most would say that humans show non-rational behavior far more frequently than they demonstrate rational behavior. Is it necessary always to be rational in order to be intelligent? If so, are humans intelligent?

As we have seen, there have been many people, worldwide, looking for signs of extraterrestrial intelligence, so far apparently with no success. In spite of SETI why has there been no contact made? Answers vary:

1. Perhaps there has been contact made but we haven't recognized it. Or perhaps contact has occurred, but has not been officially recognized, as ufologists insist. Others maintain that humans themselves are ETs – the product of intergalactic

events that explain the very sudden and recent emergence and rise to dominance of humans, compared to other species on Earth. And what are we to make of all the many ancient stone structures and huge markings on Earth, the shape of which seems to come in focus only from great heights above Earth, heights that no Earthlings could have reached until very recently? Are these not messages to or from ETs? Are they from humans, or perhaps from the ETs who placed humans on Earth?

2. Others conclude that with no contact with ETs so far, perhaps we indeed are alone, concluding either that it is therefore our duty to bring life and intelligence to a sterile universe or – especially given our planet-eating proclivity on Earth – that we should stay home and not boldly go, polluting the universe.
3. Still others conclude that ETs exist, but they choose to be silent for any one of many reasons, including the possibility that we are not yet evolved enough to be of interest to them, or they are waiting until our growing obesity makes us more appealing for eating. Some scientists use this fear to urge that we do our best not to make contact with ET since the impact of highly advanced beings will be largely negative, regardless of their intentions, which may or may not be honorable from a human point of view.
4. Perhaps ETs are deaf and blind and cannot hear or see us, or our signals, and/or we are looking/listening to/sending the wrong signals ourselves.
5. Perhaps they have already sent us many messages but that we are unable to perceive much less to understand their messages.
6. On the other hand, perhaps the universe itself is alive and intelligent, and we humans on Earth are merely barely functional players in the vast intelligence of the cosmos.

Contact!

What if we do detect a signal from or actually make contact with what seems to be ET? At the present time, in spite of considerable discussion among people in the SETI community, as we have seen, there is no agreement and no official protocol at all, so all hell might well break out. Here are some of the issues that might be important.

1. It depends on when and where a signal/contact is perceived: Is it a matter of great distance (i.e., is the signal coming from many light-years to light-centuries away before ETs or forces produced by them can impact Earth?) Is it imminent? (Is the signal only a few days, or a few years, or a few centuries away from Earth)? Is it immediate? (i.e., Will there be a physical landing on Earth or Moon, or in orbit? Is the physical object the ETs themselves, or an artifact from the ETs? Can we tell the difference?). Who/where is it received? (i.e., In the United States, China, France, Iraq, North Korea, the UN, the military, etc.?).
2. It depends on the message: Can we understand it? Do we believe it? (i.e., has someone cried wolf too many times before?) Does it seem to be clearly friendly or hostile? Can we trust it?

3. It depends on what ET does if there is contact or a physical/energy presence: Do they disrupt everything in some unique way? Do billions die or transform, or does life pretty much go on as usual?
4. It depends on our preparation: Are there agreed-upon protocols? Is the public aware of the protocols? Are they transparent and public or hidden and secret? Are the military or scientists in control?
5. It depends on what is happening when contact is made: Are we in the middle of a war, a natural disaster, or during "normal" conditions? There very well may be initial panic, but it is highly likely that will quickly pass and most people will lose interest and resume their normal lives, just adding the presence of ETs to it. It is not likely contact "will change everything" as some people say. Most religions will undoubtedly adapt and survive without missing a beat.

However, in the meantime we can just sit and wonder, "if they are, where are they"?

References

D. Atri et al., A protocol for messaging to extraterrestrial intelligence. Space Policy **27**(3), 165–169 (2011)

C. Bertka, *Exploring the Origin, Extent, and Future of Life. Philosophical, Ethical, and Theological Perspectives* (Cambridge University Press, Cambridge, 2009)

S. Dick (ed.), *Many Worlds: The New Universe, Extraterrestrial Life, and the Theological Implications* (Templeton Foundation Press, Philadelphia, 2000)

J. Pass, The sociology of SETI: An astrosociological perspective (2005), Astrosociology.com

H.P. Shuch, *Searching for Extraterrestrial Intelligence* (Springer, New York, 2010)

D. Swift, *SETI Pioneers* (University of Arizona Press, Tucson, 1990)

J. Tarter, The search for extraterrestrial intelligence (SETI). Annu. Rev. Astron. Astrophys. **39**, 511–548 (2001)

A. Tough (ed.), *When SETI Succeeds* (Foundation for the Future, Billingham, 2000)

D. Vakoch, A. Harrison, *Civilizations Beyond Earth: Extraterrestrial Life and Society* (Berghahn Books, New York, 2011)

Chapter 10
Futures of Space

It has become almost a ritual to have the last chapter of any book or the last chapter of any essay to be about the future of whatever the topic has been. It seems that is the case here as well. But careful consideration was given to having this chapter be the first – to encourage us look at everything that is and has been part of space ideas and activities through the lens of the "futures." Clearly the futures of space activities is what really matter. What is past is prologue. We need to know our past not only so we may honor and learn from the people who have labored before us – "remembering that once it was all that was humanly possible" – but also to help us understand why we feel the way we do – and don't – feel about the opportunities and challenges yet to come.

"Nowism"

Though humans are influenced by the past (or their images of the past) they are even more profoundly rooted in the now. We are biologically, psychologically, and sociologically constructed to respond to the present; to be able to adapt to whatever we encounter. This stance worked extremely well for tens of thousands of years when the past, present, and future were basically the same. In olden days whatever had worked before, and worked now, would probably work well in the future. In essence the best way to look forward in the past was to look backward and do as we were told to do by our elders. But, as we have seen repeatedly in all aspects of space activities, times have changed profoundly. We now have technologically – and institutionally – augmented powers and responsibilities that generations before us simply did not have. In today's world it is up to us to use these powers wisely and ethically. We live in environments of unprecedentedly rapid social and environmental change primarily caused by the technologies that were developed and used in the past. But we also are inventing, developing, diffusing, and using new technologies at an even faster pace, causing ever more rapid and unpredictable social and environmental changes that must be dealt with not only by our older selves, but especially by many future generations yet to come.

Nonetheless, because of our "nowism," in spite of all the change going on around us, we tend to view the future as more or less an extension of the past and present. Many of us live in the shadow of or hope for "progress" through "economic development." This indeed is the "now" for a large part of the world, especially of that part likely to be active in space ventures. It is appealing to imagine that just as the world has "advanced" from hunting and gathering to agricultural to industrial and now to information societies, and from small face-to-face homogeneous communities to global, faceless and heterogeneous institutions, that we are destined to continue to grow "forward." Those who most believe in "manifest progress" see a journey forward not just globally but throughout the inner Solar System, and eventually beyond. We will do this, we believe, in ways that will make life better for everyone, by making us more numerous, more wealthy, more mobile, more free, more individualistic. That has been the story of "progress" so far, and many of us are fully imbued with it.

Frontier Visions

Many stories told about the future of space activities are versions of the story of the frontier, of what is often said to be the fundamental fact that humans are by nature restless explorers who must always move forward towards the next frontier, and then restlessly to the next frontier after that. Ever since President John F. Kennedy made his famous declaration of landing people on the Moon, our space policies and declarations have been full of similar or even more expansive visions of the wonders yet to come. Getting off the planet and into space is said to be our manifest destiny; something that we must do, and will do, if not now, then later. And yet we have seen that many cultures do not seek out new frontiers to explore. They stay at home and grow more exotic flowers and fruit in their own garden. Indeed, the cowboy mentality of the endless frontier may be a pathology, a stage of adolescent arrested development that has characterized the two nations who initially began the space race but does not seem to be the motivation for space activities in many others.

Science Visions

Instead, space is seen by many other enthusiasts as a part of science, and not of politics and economics. Their enthusiasm is driven by the desire to know, not the desire to conquer and exploit. And, like all science, it needs to be done carefully, slowly, repeatedly, with many frequent corrections and restatements. We should not boldly go, but humbly go, and only go when we are fully ready for everything the journey entails, including a robust sense of ethical responsibility towards the life and environments we will encounter. To many, the origins of the space age – the space race to the Moon – was a tragic mistake, undertaken many decades, if not centuries, before humans were ready for it. And so it is no surprise to them that we have not gone back to the Moon, much less elsewhere, and don't seem about to go anywhere now either, in spite of reams of declarations to the contrary. Indeed, while we have largely "gone to space" by simply going round and round Earth, even the continuation of that seems very problematic. The era of space exploration will be over, at least for a while, and perhaps forever, if we run out of sufficient cheap and abundant energy sources, as seems possible, or pollute the environment of near-Earth space with so much junk that it will be physically impossible for humans ever to leave.

What's the Rush?

If the space age might well be over less than a century after it was begun, then many other people say we need to consider our options very, very carefully, and not waste our dwindling resources and talents on continued national or commercial competition. Some say we need to make space a human, global, scientific activity, and not a competitive political and/or economic activity. We should take our time to get it right. We must make this a major activity of scientists, engineers, social scientists, humanists, philosophers, and artists all over the world – the major long-range project for humanity over the next centuries. Take it easy. Get it right. Take care of things on Earth, but set your sights for the stars and we will get there in due time.

Boldly Go!

Bullfeathers! say others who are appalled by such timidity and shortsightedness. Just look at the developments that are being made and will continue to be made in new energy and materials, and especially in the development of electronics, biology, and nanotechnology which, among other things, clearly are leading to the end of the dominance of old *homo sapiens*, and the emergence of post-humans, transhumans, artilects, cyborgs and many other new forms of life and intelligence. While the old space programs required enormous amounts of energy, huge rocketships, and extremely expensive systems to keep humans alive, Freeman Dyson's old dream of humans moving into space on the wings of a space butterfly or an astrochicken might finally come true. The former he described as "a way of exploiting for the purposes of space science the biological technology which allows a humble caterpillar to wrap itself up in a chrysalis and emerge 3 weeks later transformed into a shimmering beauty of legs and antennae and wings." "So it is reasonable to think of the microspacecraft of the year 2010 not as a structure of metal and glass and silicon but as a living creature, fed on Earth like a caterpillar, launched into space like a chrysalis, riding a laser beam into orbit, and metamorphosing itself in space like a butterfly." Alternatively, Dyson says the "Astrochicken will not be built, it will be grown. It will be organized biologically and its blueprints will be written in the convenient digital language of DNA." "The next hundred years," says Dyson, "will be a period of transition between the metal-and-silicon technology of today and the enzyme-and nerve technology of tomorrow. The enzyme-and-nerve technology will be the result of combining the tools of genetic engineering and artificial intelligence."

Dyson was obviously off quite a bit in his timing, but many adhere to his dream, urging us to throw off the shackles of what was technologically and sociologically necessary in the 1950s and 1960s and embrace what may be possible for the 2050s and 2060s – and certainly later.

Four Alternative Futures

It is impossible to "predict" the future – to say with accuracy what WILL happen. That might have been possible once upon a time, long ago, but not now. What we can do is to "forecast" alternative futures. In fact, there are always four alternative futures ahead of us. They can be labeled "Grow," "Collapse," "Discipline," and "Transform." We have used that template in the paragraphs above, suggesting different alternative futures for space. We need to contemplate each of the four as fully and seriously as we can, and then try to envision, invent, and move towards our "preferred future," which emerges from them.

A preferred future is not a utopia. It is a "eutopia." Utopia means "no place" and exhibits the dangerous allure of an impossibly perfect future. Eutopia means "good place," and so a "preferred future" is the best possible real world we can imagine, given the forces pushing us from the past and pulling us towards the future. Most importantly, we must continuously monitor the world around us and ahead, and constantly modify our preferred future as new people are born and old ones die, new technologies emerge and old ones fade, and new challenges confront us as we overcome old ones. A preferred future is an image, a compelling, convincing image that is constantly changing as events and opportunities change.

A Preferred Future Exploring Not-Earth

For many people, including this author, one preferred future is very much along the "transformational" lines of Dyson vision, and as also expressed by University of Hawaii anthropologist Ben Finney. Finney's expertise is the experience of the Pacific voyagers who left southern Asia 2,000 years ago and gradually spread out in outrigger canoes until they had occupied every possible speck of land in the Pacific from what is now called Melanesia in the southwest, to Polynesia from Aotearoa (New Zealand) in the south to Hawaii in the north, and to Micronesia in the western Pacific. What was one small community of people with a single language and culture developed, as a consequence of voyaging so far and so long and then living in very different environments almost totally separated from other environments, into many different cultures with mutually unintelligible languages and people with very different and distinctive physical features.

Finney sees this as an excellent analogy for what may happen once humans leave their cradle, Earth, and begin to live within the Solar System and beyond: "If our descendants spread far and wide through space, the forces of evolution now braked on Earth will be released once more .... Human evolution in space will hardly be limited to the birth of one new species. Space is not a single environment .... There are innumerable environments out there providing countless niches to exploit, first by humans and then by the multitudinous descendant species. By expanding through space we will be embarking on an adventure that will spread an explosive speciation of intelligent life as far as technology or limits placed by competing life forms originating elsewhere will allow."

If the life and intelligence of cyborgs, artilects, and trans-humans are added to this mix of "naturally" evolving post-homo sapiens, the future of spacekind as it evolves from Earthkind, as Jerry Glenn termed it long ago, seems bright indeed.

However, we will never achieve this future until we solve looming energy shortages on the one hand and refrain from ever-more wasteful wars on the other, both of which seem very doubtful.

What is your preferred future for Earth and not-Earth? How can you help achieve it?

References

W. Bell, *Foundations of Futures Studies: Human Science for a New Era* (Transaction Publishers, New Brunswick, 1997), two volumes

B. Cordell, 21st century waves: forecasting technology booms and human expansion into the cosmos. Futures Res. Q. **22**(3), 21–41 (2006)

J. Dator (ed.), *Advancing Futures: Futures Studies in Higher Education* (Praeger, New York, 2002)

J. Dator, Futures studies and futures research, in *Leadership in Science and Technology*, ed. by W.S. Bainbridge (Sage, Thousand Oaks, 2011), Reference Series

B. Finney, E.M. Jones (eds.), *Interstellar Migration and the Human Experience* (University of California Press, Berkeley, 1985)

G. Genta, M. Rycroft, Will space actually be the final frontier of humankind? Acta Astronaut. **58**, 287–295 (2006)

W. Halal, Forecasting the future in space: highlights from the TechCast Project. Space Policy **23**, 234–240 (2007)

P. Harris, *Space Enterprise: Living and Working Offworld in the 21st Century* (Springer Praxis Books, New York, 2009)

R. Lanius, H.E. McCurdy, *Imagining Space: Achievements, Possibilities, Projections, 1950–2050* (Chronicle Books, San Francisco, 2001)

R. Launius, Perfect worlds, perfect societies: the persistent goal of Utopia in human spaceflight. J. Br. Interpla. Soc. **56**, 338–349 (2003)

R. Launius, D. Jenkins, Is it finally time for space tourism? Astropolitics Int. J. Space Politics Policy **4**, 253–280 (2006)

C. Mari (ed.), *The Next Space Age* (H. W. Wilson, New York, 2008)

W. Mendell, Meditations on the new space vision: the Moon as a stepping stone to Mars. Acta Astronaut. **57**, 676–683 (2005)

F. Ordway III, R. Liebermann (eds.), *Blueprint for Space* (Smithsonian Institution Press, Washington, DC, 1992)

W. Peeters, From suborbital space tourism to commercial personal spaceflight. Acta Astronaut. **66**, 1625–1632 (2010)

K.S. Polk, *Gaiome: Notes on Ecology, Space Travel and Becoming Cosmic Species*. (gaiome.com, Bloomington, 2007)

C. Sagan, *Pale Blue Dot: A Vision of the Human Future in Space* (Random House, New York, 1994)

R. Zubrin, *Entering Space: Creating a Spacefaring Civilization* (Tarcher/Putnam, New York, 2000)

Index

J.A. Dator, *Social Foundations of Human Space Exploration*, SpringerBriefs in Space Development, DOI 10.1007/978-1-4614-3094-0, © James A. Dator 2012

MIX
Papier aus verantwortungsvollen Quellen
Paper from responsible sources
FSC® C105338

FSC
www.fsc.org

If you have any concerns about our products,
you can contact us on
ProductSafety@springernature.com

In case Publisher is established outside the EU,
the EU authorized representative is:
Springer Nature Customer Service Center GmbH
Europaplatz 3, 69115 Heidelberg, Germany

Printed by Libri Plureos GmbH
in Hamburg, Germany